Yes, the world faces substantial environmental challenges—climate change, pollution, and extinction. But the surprisingly good news is that a remarkable number of environmental problems have been solved, while substantial progress is ongoing on others.

The Optimistic Environmentalist chronicles these remarkable success stories and suggests a bright green future is not only possible, it's within our grasp.

The
Optimistic
Environmentalist

Progressing Towards a Greener Future

David R. Boyd

ECW Press

Also by David R. Boyd

Cleaner, Greener, Healthier:
A Prescription for Stronger Canadian Environmental Laws and Policies (2015)

The Right to a Healthy Environment: Revitalizing Canada's Constitution (2013)

The Environmental Rights Revolution:
A Global Study of Constitutions, Human Rights, and the Environment (2012)

Dodging the Toxic Bullet:
How to Protect Yourself from Everyday Environmental Health Hazards (2010)

David Suzuki's Green Guide (with David Suzuki, 2008)

Unnatural Law: Rethinking Canadian Environmental Law and Policy (2003)

For Meredith and Margot

Optimistic: "hopeful and confident about the future"
—Oxford English Dictionary

"Another world is not only possible, she is on her way. On a quiet day, I can hear her breathing."
—Arundhati Roy

"Most of the things worth doing in the world had been declared impossible before they were done."
—Louis D. Brandeis

Contents

Introduction:
The Importance of Being Optimistic   xi

Part I: THE BIG PICTURE
1 | Nature's Comeback Stories   3
2 | The Renewable Energy Revolution   25
3 | The Circular Economy   51

Part II: HEALTHY ENVIRONMENT, HEALTHY PEOPLE
4 | Cleaner Air   71
5 | Saving the Ozone Layer   87
6 | Taps, Toilets, and Farms   105
7 | Global Detox   123

Part III: THE BUILT ENVIRONMENT
8 | The Greenest City Decathlon   143
9 | The Future of Buildings   159
10 | Electrifying Transport   173

Conclusion:
From Optimism to Action   187

Selected Bibliography   205
Acknowledgements   217
About the Author   219

Introduction
The Importance of Being Optimistic

I'M AN OPTIMISTIC ENVIRONMENTALIST. Unlike open secret, jumbo shrimp, or working vacation, that's not an oxymoron. But I want to be crystal clear: our society faces serious environmental challenges, including climate change, toxic pollution, and the declining diversity and abundance of plant and wildlife species. The scientific evidence is irrefutable. But, based on humanity's track record over the past 50 years, the ready availability of effective solutions, and the potential of future innovations, I also believe that today's environmental challenges can be overcome. From air pollution to safe drinking water, from greener cities to renewable energy, we've made remarkable but widely underacknowledged progress. If you feel overwhelmed or exhausted by the onslaught of bad news about the planet and are looking for some genuinely good environmental news, then you've come to the right place.

I'm not the "bury your head in the sand and pretend everything is hunky-dory" type like Danish statistician Bjorn Lomborg or American writer Matt Ridley. Lomborg is a 21st-century snake oil salesman who has made a lucrative career out of downplaying the world's environmental challenges. His 2001 book, *The Skeptical Environmentalist*, manipulated data to suggest that scientists and activists had systematically concocted or inflated every environmental problem in the world. Ridley followed suit in 2010 with *The Rational Optimist*, arguing that the dangers of acid rain, falling sperm counts, the ozone hole, and desertification are nothing but "eco-exaggeration." But why would almost all of the world's millions of scientists engage in such a complex and convoluted conspiracy? The answer is beyond my imagination. Despite their absurd arguments, Lomborg's and Ridley's books have been bestsellers. Apparently there's a large appetite for good news, even when the underlying premises are false.

I'D BEEN THINKING ABOUT WRITING a book like this one for years, but my daughter, Meredith, provided the spark that propelled me to action. At the wise old age of seven, Meredith came home from school one afternoon and told me about her day. She'd learned about something called global warming and described it to me like this: "Pollution is melting the ice in the Arctic Ocean. Habitats are disappearing, and the animals don't have anywhere to live. Polar bears are dying." Tears welled up in her big blue eyes. She's had a soft spot for polar bears since getting a big stuffed one as a gift from her Grandma Grace. "Species are going extinct," Meredith said, beginning to cry.

As kids do, she watched my reaction closely, looking for subtle signals in my body language or facial expression. I'd been dreading this moment. Having worked as an environmental lawyer for more than 20 years, I was painfully aware of the world's converging eco-crises. There's climate change—rising sea levels, super-storms, droughts, and ocean acidification—caused by our reckless burning of fossil fuels, clearing of tropical rainforests, and industrial farming methods. There's the planet's sixth mass extinction, the most devastating since *Tyrannosaurus rex* and other dinosaurs disappeared 65 million years ago. The bodies of everyone living in the industrialized world carry a toxic burden of plastic, pesticides, and hundreds of other industrial chemicals. There's even a new word, Anthropocene, signifying the end of the Holocene epoch and the beginning of a new geological era in which a single species, *Homo sapiens*, has planet-wide ecological impacts. The sheer volume of bad news about the environment can be overwhelming.

I was torn between conflicting instincts. On the one hand, I value honesty and steadfastly avoid lying to Meredith. On the other hand, I want to shelter my daughter from the worst of the world's news until she has the emotional maturity and intellectual ability to cope with these issues. So very carefully I said, "It's true that humans are causing environmental problems. But we're also pretty good at solving those problems, and millions of people all over the world are doing their very best to help prevent pollution and extinction." Searching for an example that she could understand, I told Meredith the story of sea otters, a charismatic creature that's recovering nicely from a human-induced brush with oblivion.

First I had to clarify the difference between sea otters and river otters, because we only see the latter where we live on Pender Island, between Vancouver and Victoria on Canada's west coast. River otters forage for food in the ocean but make their homes on land, unlike their marine cousins. They're smaller than sea otters and I don't think they're as cute, though that opinion comes from someone who has involuntarily shared his house with these malodorous creatures. Once when Meredith was a toddler, we were out on the deck at night looking at constellations. A pair of otters began mating loudly and enthusiastically below us, leading Meredith to say, "Papa, the stars are singing." River otters loved spending parts of winter in the warm, dry crawl space beneath our house, covering it with a ghastly combination of shell-filled excrement and a diabolically putrid mucous-like excretion used to mark their territory. They were unbelievably bad housemates. For years, the otters outsmarted our attempts to evict them. Several years ago, we had to build a concrete perimeter foundation as part of a major home repair project, and that has finally kept them out.

Sea otters, happily for human residents of the west coast, spend their entire lives in the ocean, diving to the sea floor to find marine invertebrates like crabs, clams, and sea urchins. They float on their backs, eating and resting in rafts that usually number 10–100 animals, though super-rafts as large as 2,000 have been reported. Only rarely do they venture onto land, and they'll never invade your home. Sea otters were long targeted for their luxurious fur, hunted at sustainable levels by Indigenous people prior to the arrival of Europeans, and then in a ferocious fur-trading frenzy that didn't stop until there were no sea otters left on the Canadian coast. Despite an international protection treaty negotiated in 1911, the last Canadian sea otter was killed near Kyuquot, on Vancouver Island, in 1929. The global population was slashed from between 150,000–300,000 sea otters to 1,000–2,000 individuals. In 1978, shortly after the Committee on the Status of Endangered Wildlife in Canada was created, sea otters were among the first species in Canada to be officially designated as endangered.

I'd never understood the attraction of fur, until one day I went behind the scenes at the Vancouver Aquarium, viewing items that are not put out for public display. I almost passed on the opportunity to handle a sea otter pelt, based on a lifelong aversion to the idea of killing animals for their skins. But the fur was handed to me by a young aquarium employee, who explained that sea otters have up to a million hairs per square inch, a mind-boggling figure. As I stroked the fur, I was shocked to find atavistic desire flitting across the edge of my consciousness. It was so soft! Some primal part of me didn't want to give it back. The sea otter's coat compares to a total of 20,000 hairs on the average human head (and far fewer on mine). Even domestic cats have relatively sparse coats compared to those of sea otters, with 60,000 hairs per square inch on their backs and up to 120,000 on their bellies. In other words, even your favourite feline, whose soft coat you love to pet (when it permits you to do so), is effectively bald compared to the sea otter, a feline Bruce Willis to the otter's Troy Polamalu.

Fortunately, some sea otters survived in Alaska. After an absence of half a century, Canada asked to borrow a few animals to start a recovery program. Between 1969 and 1972, 89 sea otters were flown or shipped from Alaska to Vancouver Island. Meanwhile, scientists were learning about the critical role that sea otters play in the marine ecosystems of North America's west coast. Sea otters are a keystone species whose absence threw entire systems out of whack. Otters eat up to one-quarter of their body weight daily, and one of their favourite foods is the purple sea urchin. When sea otters disappeared, the sea urchins went wild, munching their way through kelp beds. Those kelp beds, it turns out, provide critical habitat and nutrients for various species of fish. No habitat, no nutrients, no fish. This had a domino effect on other species, creating areas called urchin barrens. Of course Nature is always evolving, but the absence of sea otters created a cascade of unanticipated ecological impacts that diminished the biodiversity of the coast.

The *Exxon Valdez* oil spill in 1989 killed thousands of sea otters in Alaska, caused elevated mortality rates for years after the spill as otters continued to be exposed to high levels of hydrocarbons, and had long-term impacts on sea otter populations. Concerns about a similar disaster are a leading reason for the strong opposition, shared by the majority of British Columbians, to proposed pipelines that would bring additional tanker traffic to the coast.

The 89 imported sea otters have flourished in British Columbia's coastal waters, reproducing successfully, spreading north and south, and re-inhabiting regions where they were once abundant. The species has bounced back so strongly that scientists down-listed its status from endangered to threatened in 1996 and from threatened to special concern in 2007 (the lowest category in Canada's hierarchy of species at risk).

Meredith was satisfied by my explanation, but I realized the questions she'd raised that day were just the beginning. Today's children grow up in a wired world where bad news about the environment is ubiquitous. Catastrophes like BP's massive oil spill in the Gulf of Mexico garner saturation coverage. Yet just as headline news about shocking murders or plane crashes obscures the truth about falling rates of violent crime and the impressive safety record of airlines, media coverage of environmental issues focuses inordinately on bad news, overlooking stories about progress.

Ditto for popular culture. Judging by the surge of dystopian novels, films, and video games, you'd be forgiven for thinking that the end is nigh. Young adult fiction is brimming with dark postapocalyptic epics such as *The Hunger Games*, *Divergent*, *Ship Breaker*, and *The Drowned Cities*. Hollywood is cranking out doomsday movies like *Oblivion*, *The Purge*, and *Dawn of the Planet of the Apes*. Even *Wall-E*, an animated film intended for young children, features a planet rendered uninhabitable by pollution and overconsumption. Some of my favourite novels are set in bleak futures after awful environmental catastrophes, including Margaret Atwood's recent trilogy (*Oryx and Crake*, *After the Flood*, and *MaddAddam*), P.D. James' *The Children of Men*, Russell Hoban's *Riddley Walker*, Ronald Wright's *A Scientific Romance*, and Cormac McCarthy's *The Road*.

All of this negativity inflicts real harm. After all, human beings are hard-wired for optimism—neuroscientists believe it was selected for success through the trials and tribulations of evolution. Humans needed to be optimistic to venture forth from Africa 150,000 years ago in search of new homelands. Humans needed to be optimistic to plant the first seeds of the agricultural era, believing that putting down roots in one place would be rewarded eventually. We need to be optimistic to have children, making tremendous investments of time, energy, and money in the belief that their lives will be as good as, or better, than our own. And we need to be optimistic about our prospects for becoming better environmental stewards, so that we'll take the actions required to save the planet.

Thus optimism is more than just the kind of sunny disposition that makes a person pleasant company. Optimism is a powerful causal factor in shaping outcomes and futures. Three fascinating studies published recently revealed that optimists are more likely to recover from cancer, limit the adverse effects of chronic illnesses such as Parkinson's disease, and cope successfully with the effects of aging. American swimmer Matt Biondi was part of a study examining the relationship between optimism and athletic performance. At the 1988 Olympics in Seoul, Korea, Biondi was widely expected to win seven gold medals, but in his first two races won silver and bronze. Psychologists credit Biondi's optimism for enabling him to bounce back from those disappointments and capture five consecutive gold medals. Overall, optimists are happier, healthier, and more persistent in overcoming obstacles and achieving their goals.

Pessimistic people are worse off than optimists in many ways—more vulnerable to mental illness; less successful in school, work, and sports; and prone to worse physical health. When adverse events occur, pessimists are eight times more likely to suffer depression. Compared to optimists, pessimists set lower expectations for themselves, experience less satisfying personal relationships, and quit trying earlier. As Winston Churchill wrote, "The pessimist sees difficulty in every opportunity. The optimist sees opportunity in every difficulty."

Despite growing scientific recognition of the importance of optimism, we live in a society haunted by an epidemic of depression. It's the second leading cause of disability in the world, an invisible pandemic that harms many millions of people. Worse yet, growing numbers of children are afflicted. One in three teenagers in the U.S. has shown symptoms of depression, while one in six has suffered from an episode of major depression. Relentlessly negative messages about humans destroying the fragile blue-green planet that we call home are a significant contributing factor.

Our dispositions determine how we view the world. Several years ago, after decades of absence, humpback whales began returning to the waters of the Salish Sea. My optimistic friends and I saw the reappearance of humpback whales as a powerful good news story, signaling cleaner water and recovering marine ecosystems. The pessimists' reaction to the return of the whales was diametrically opposite—despair instead of joy. From their perspective, these whales were fleeing a disaster elsewhere, driven into our contaminated and depleted waters by even worse pollution and ecological collapse.

Research by psychologists and cognitive linguists offers profound insights for those who care about the planet's future. Pressing people's fear buttons, a prominent strategy among environmentalists, usually triggers an instinctive survival response. When confronted by fear, most people suppress their concern for others and focus on their own interests. In the words of British author and journalist George Monbiot, "Terrify the living daylights out of people, and they will protect themselves at the expense of others and the living world." In the face of overwhelming environmental threats, people are overcome by feelings of helplessness and are less likely to take any kind of remedial action. They'll even begin to deliberately avoid the issue. In contrast, hope is like an elixir for action.

ALTHOUGH I'M GENERALLY a positive person and am now a passionate optimist, I have to make an uncomfortable confession. I've personally experienced the onset of the black dog, as Churchill referred to depression. Only a handful of family and friends know this, as I've managed to maintain a cheerful outward veneer. Discouraging news stories about the environment triggered my descent into darkness. As the great American wilderness writer Aldo Leopold observed, "One of the penalties of an ecological education is that one lives alone in a world of wounds."

I became an environmental lawyer in 1993, at a time of great excitement about our collective ability to solve environmental problems. Nations had recently tackled the terrible threat posed by the thinning of the planet's ozone layer. Canada hosted the meeting in 1987 that resulted in the widely praised Montreal Protocol on Substances That Deplete the Ozone Layer. The Rio Earth Summit in 1992 reflected a world united by unprecedented environmental concern. All of the world's presidents and prime ministers attended the Earth Summit, working long hours to find common ground and identify solutions. New international treaties were negotiated, promising to address climate change and reverse the decline of biodiversity. Canada was the first industrialized country to sign the Convention on Biological Diversity and the Framework Convention on Climate Change. With my legal education, I was confident and excited about contributing to further progress. I landed my dream job with the Sierra Legal Defence Fund (now Ecojustice Canada), a new organization dedicated to using the courts to win environmental victories.

After a decade of practising environmental law, my optimism lay in tatters. Suing governments and corporations for breaking environmental laws didn't seem to produce much progress, even when we won big cases at the Supreme Court of Canada. My beloved country was backsliding badly, and studies (some of which

I authored) showed that Canada's green record was among the worst in the industrialized world. I wrote a book called *Unnatural Law: Rethinking Canadian Environmental Law and Policy* which chronicled, in painful detail, the legal system's role in Canada's weak environmental performance. These experiences led to the deepest depression of my life.

To rebuild a positive attitude, my focus shifted to teaching environmental law to bright, committed young people and studying other countries to learn about successful laws, policies, and approaches. I did a Ph.D. on the effects of enshrining constitutional protection to the right to live in a healthy environment and was delighted to discover its surprisingly widespread and positive impacts. This led to work with countries from Iceland to Tunisia, crafting strong green provisions for their new constitutions. Now I teach judges from all over the world about the vital relationship between human rights and the environment, receiving a much warmer reception from them in the classroom than I ever did in the courtroom. I also worked with the government of Sweden to strengthen their world-leading effort to achieve sustainability within a generation (see Conclusion). Mercifully, my depression ended and has not returned. The arrival of our daughter Meredith in 2006 has been a constant source of joy and inspiration, reinforcing my renewed sense of hope.

I'M WRITING THIS BOOK in an eight-by-twelve-foot cabin beside our home on North Pender Island. A single south-facing solar photovoltaic panel powers my laptop and an array of super-efficient LED lights. Even on a foggy October morning, with ferry horns blaring, the system still cranks out electricity. A couple of batteries store the sun's energy, enabling me to work late into the night when I'm on a roll. The system was expensive when it was installed in 2010. Five years later, the prices of solar panels and LED lights have plunged by up to 80 percent. Today, solar is an economical means of providing electricity, even on Canada's notoriously wet west coast. Prices for solar electricity have fallen so fast that, in a 2014 survey, 97 percent of Americans overestimated them.

Looking directly south from the cabin, I can see over top of Moresby Island through the Strait of Juan de Fuca to the snow-capped Olympic Mountains in Washington State. To the southwest, the morning sunlight reflects off a hilltop observatory in Vancouver Island. To the west is Ruckle Park on Saltspring Island. To the east, I can see Roesland, a small point and bay that is part of the southern Gulf Islands National Park. Alas, while the water looks enticing, it's a bone-chilling temperature year-round (though the area is superb for sea kayaking). Regardless of the larger woes gripping the world, Pender Island is a refuge, a lovely place to raise a child, lead a quiet and contemplative life, and write books.

It would have been fascinating to travel the world seeking out stories for this book but I didn't have the budget (either financial or carbon). Instead of flying, I visited solar and geothermal power plants, cities like Copenhagen and Amsterdam, and state-of-the-art wastewater treatment plants virtually, going on tours via YouTube and Google Earth. I interviewed far-flung experts using Skype. Since Pender Island is just a short ferry ride from Victoria and Vancouver, I was able to visit super-efficient passive houses and living buildings, cycle safely in separated bike lanes, and take a Tesla Model S electric car for a test drive. On a summer holiday, I visited my cousins' innovative organic farm in Alberta. Finally, I drew on experience garnered while visiting places from Scandinavia to South America to South Africa for my work on environmental policy, constitutions, and human rights.

My goals in writing this book are to provide an antidote to the plague of ecological negativity; to give people a sense of optimism that a greener, cleaner, healthier, and happier future is possible; and to inspire people to act. A bright future isn't a 99 to 1 long shot, but rather the odds-on favourite. As you'll see, my hopeful attitude isn't based on cherry-picking a few heart-warming anecdotes—students cleaning up a local beach, or a progressive government creating a park to protect a particularly beautiful place. Instead, my conviction that a sustainable and prosperous future lies within our reach is based on a sober and balanced examination of the facts about humanity's past environmental successes, current trends, and future probabilities. The extent of the good news that I discovered while researching this book astonished me.

Dedicated and sometimes heroic efforts to protect the environment over the past 50 years have produced a surprisingly long list of victories and success stories. Endangered species—from sea otters and bald eagles to gray whales and black-footed ferrets—have been painstakingly pulled back from the precipice of extinction. Tens of thousands of new parks and ecological reserves have been established across the globe, protecting billions of hectares of land and water. Deforestation is declining, and in many countries forests are rebounding from historic lows. Human ingenuity inadvertently threatened the vital ozone layer, which protects life on Earth from UVB radiation emitted by the sun, but when scientists discovered and described the magnitude of the threat, people and politicians responded swiftly and effectively, phasing out the use of CFCs and other chemicals that caused the problem. Clean, renewable energy harnessing wind, water, and most importantly the sun is growing at an exponential rate. Across the planet, municipalities are competing in a race to be the greenest city in the world. Transportation patterns show the rising popularity of cycling, walking, public transit, and zero-emission vehicles, while innovations such as self-driving cars, high speed trains in vacuum tubes, and solar roads lie on the horizon, tantalizingly close. Buildings are being designed and built to produce more energy than they consume.

There's more. We've made remarkable progress in reducing air and water pollution. Billions of people are breathing cleaner air, drinking safe water, and enjoying unprecedented access to modern sanitation infrastructure. Levels of some air pollutants are down 90 percent in cities across North America, Europe, and Australia. The leaded gasoline that damaged millions of children's brains during the 20th century has been phased out in almost every country. We've eliminated the production, use, and release of dozens of the world's most toxic chemicals, from pesticides to flame retardants. More are being phased out. Levels of these contaminants in the environment and our bodies are declining. We've dramatically reduced the acid rain that ravaged lakes, forests, and soils. Organic agriculture and local food are wildly popular. A growing number of companies and countries are working to create a circular economy, making concepts such as waste and pollution obsolete.

Over the past 50 years, we've also witnessed an extraordinary transformation of legal systems and the rules that govern our societies. Hundreds of international environmental treaties have been negotiated and ratified. Environmental laws have been enacted in every country. Agencies responsible for environmental protection have been created at every level of government in every nation. A new human right—to live in a healthy environment—has emerged and is now endorsed by 90 percent of the world's countries. This fundamental right is protected in over 110 constitutions, making it not only legally powerful but also indicating that it's among our deepest and most cherished values.

By exploring these unsung environmental success stories, this book will illustrate our ability to achieve the future that so many people yearn to believe in but have difficulty imagining. A future where the human impact on the planet is reduced to sustainable levels and the ravaged ecosystems of today are restored to flourishing levels of diversity and abundance. A future where even large cities are sparkling green jewels. A future where we can all look our children in the eye and honestly say, "We did the best we could. For the most part, we've solved the environmental problems that we created. Please learn from our mistakes and emulate our successes. The Earth is a beautiful planet to call home."

Part I
THE BIG PICTURE

"How wonderful it is that nobody need wait a single moment before starting to improve the world."
—Anne Frank

Chapter 1
Nature's Comeback Stories

THE MEDIA REGULARLY REPORTS heart-rending stories about species pushed to the brink of extinction by human malfeasance—overhunting, overfishing, destroying habitat, introducing alien species, and spewing toxic substances into the environment. It's true that the rate of extinctions has accelerated in recent centuries. Despite this, many species are enjoying remarkable comebacks because we've smartened up and improved our once-damaging ways.

One of the first memorable slogans of the environmental movement in the early 1970s was "Save the Whales." Who can forget the iconic image of the first Greenpeace activists in a tiny Zodiac, buzzing around a Russian whaling vessel like an agitated bumblebee trying to protect its honey from a bear? You don't hear about saving the whales that often anymore, because many whale species are making extraordinary comebacks.

From my little writing cabin overlooking Swanson Channel in the Southern Gulf Islands, I can sometimes hear whales passing by. Most of the time it's a pod of southern resident killer whales, hot on the trail of a school of Chinook salmon. In recent years, humpback whales have reappeared. Not in huge numbers, and they can't be sighted daily, but they appear with a frequency and consistency that is encouraging. As with the more commonly observed orcas, you hear them before you see them. When humpbacks surface, they exhale, a frothy whooshing blast of air that sounds like someone trying to play a waterlogged tuba. The first time my daughter Meredith heard the telltale whoosh, she thought it sounded like a sea monster. We saw the tail flukes wave at us as the whale submerged and then we watched as it surfaced and submerged repeatedly, slowly moving away to the east.

In 2014, the government of Canada announced that it was down-listing the Pacific Ocean population of humpback whales from threatened to special concern. Great news, right? Instead of prompting celebrations about the recovery of a previously imperiled creature, the news provoked criticism and controversy. Environmentalists accused the government of down-listing humpbacks in order to smooth the waters for the proposed Northern Gateway project, which involves a new pipeline from northern Alberta's bitumen sands to Kitimat on B.C.'s coast. From there, heavy crude would be loaded onto massive tankers, then navigated through a treacherous stretch of water where the *Queen of the North* ferry sank in 2006 and onwards to oil-thirsty consumers from California to China. The tanker route is a concern because it would pass directly through one of four areas identified by scientists as critical habitat for humpback whales.

Our predecessors treated humpbacks and other whales as nothing more than an infinite supply of natural resources, greedily hunting them for the oil their bodies contained. The Pacific humpback population that was once greater than 125,000 was decimated by the 1960s, with fewer than 10,000 remaining. When hunting was banned in 1966 under the auspices of a global treaty called the International Convention for the Regulation of Whaling, humpbacks must have breathed a collective whoosh of relief. Their numbers slowly began to climb and are now estimated at more than 80,000, with steady annual increases of 4–5 percent.

The humpback's cousin, the gray whale, was also nearly hunted into oblivion. Slurping sediments from the ocean's floor and filtering amphipods and shrimp through their sieve-like baleen, gray whales grow to the bus-like size of 15 metres and 35,000 kilograms. Newborns are a mere five metres and less than 1,000 kilograms but grow rapidly. If those numbers seem too abstract to give you a sense of their immense size, try this: a gray whale calf drinks between 700 and 1,100 litres of its mother's milk every day! In comparison, most human babies drink less than a litre of milk daily. Gray whales migrate from the Arctic waters of the Bering Strait to warm lagoons along the west coast of Mexico. That's among the longest migrations of any animal in the world, an annual round-trip journey of between 16,000 and 22,000 kilometres.

On one of the most wonder-filled trips we've ever taken, my wife, Margot, and I went kayaking in Mexico's Loreto Bay National Marine Park. We paddled from island to island in the Sea of Cortez on the inland coast of Baja California. The Sea of Cortez is like an enormous bowl of plankton soup, which explains why it is visited by as many as half of the world's whale species. After hearing what sounded like a volcanic eruption, we saw several blue whales, the largest living creatures on Earth. The eruption was the sound of their exhalation as they came to the surface to breathe. Several days later, a fin whale, 70 feet long, swam directly beneath our kayaks, surfacing to our right so its dinner plate–sized eye could have a good look at us. We swam with pods of playful dolphins, which are lovely to watch but suffer from extreme fish breath when you get too close.

The highlight of our trip came after taking a bus across the Baja peninsula to Puerto López Mateos, a sleepy fishing village on the edge of the Pacific Ocean where the local *pescadores* seasonally hang up their nets and switch to ecotourism. Gray whales return each spring to the saltwater lagoons to give birth. The high salinity provides a boost, literally, to newborn whale calves who easily float at the surface of these protected waters.

A dark, blood-stained shadow hangs over these lagoons: tens of thousands of gray whales were slaughtered here during the whale oil era just decades ago. The species was nicknamed the devilfish because of their violent reaction to being harpooned and fierce defence of their calves. Gray whale populations were shattered.

And yet today, the gray whales are one of the most inspiring comeback stories in the natural world. Standing on the shore at Puerto López Mateos, we could see dozens. We hired a friendly fisherman to take us for a closer look in his panga, a 16-foot boat with a small engine. After motoring across the glassy water for about ten minutes, Jorge cut the engine, leaving us to bob and drift on the gentle tidal current. Within minutes, a gray whale calf, distinctive because its skin was smooth-looking and free of barnacles, approached the boat. With the brutal history of the area swirling through my mind, I was struck by this animal's trust and innocence. The calf, glistening dark gray and the size of a Volkswagen van, swam right alongside our panga, which seemed to shrink in size. Jorge suggested we rub its head. This struck me as a bad idea. It was a wild animal, inherently unpredictable. And what about the stink of humans? Germs on our hands? What about conditioning it to approach people, not all of whom would be so friendly? Ultimately though, it was irresistible. The calf looked me in the eye and placed its head directly alongside the boat. I reached over the side and stroked its forehead. The skin was cool to the touch, but it felt like electric shocks coursed through my body. It was obviously sentient, intelligent . . . tears of joy sprang from my eyes. Then a full-sized, barnacle-encrusted gray whale, presumably its mother, gently nudged it away from the boat. We drifted around for a couple of hours on our gray whale meet-and-greet session. All the calves seemed curious and all the mothers protective. Some individual gray whales have lived 80-plus years, meaning they survived and may even remember the massacres that took place. It might sound flakey, but I felt a deep connection to those gray whales and they have left a magical and indelible memory.

In the mid-1930s, the League of Nations (the UN's short-lived predecessor) adopted a ban on commercial hunting of several whale species in recognition of a rapid decline in whale populations. This ban was the first international agreement to protect whales. It wasn't until 1982 that all commercial whaling was terminated by the International Whaling Commission, to enable populations to recover. Gray whales are still hunted by Indigenous people, subject to catch limits under the IWC's Aboriginal subsistence whaling program. Several nations continue to hunt gray whales in defiance of international law, though in far smaller numbers than in the past. In 2014, the World Court ruled that Japan's whaling program was illegal and that Japan's excuse that it was conducting scientific research was unconvincing. With limited hunting, the remaining threats—collisions with vessels, entanglement in fishing gear, noise, and pollution—do not appear to threaten the survival of the species. Gray whales were removed from the U.S. endangered species list in 1994, as the population had recovered to between 25,000 and 30,000 individuals.

Routinely visible from where I sit pecking away on the keyboard is a pair of bald eagles, perched atop a towering Douglas fir tree as though posing for *National Geographic*. Bald eagles are ubiquitous here on the south coast of British Columbia, making it difficult to imagine that when I was a young boy they nearly disappeared due to another act of human hubris. Unlike whales, it wasn't hunting that endangered the eagles, but pesticides. The main culprit was DDT, which netted Swiss chemist Paul Hermann Müeller a 1948 Nobel Prize in a classic example of premature congratulations. Between 1942 and 1972, a staggering billion pounds of DDT were used in the U.S. alone. The foolhardy application of pesticides that were never tested for potential adverse health and environmental effects prompted Rachel Carson to write her classic book *Silent Spring* (1962), warning of the deadly impact on birds. DDE, a compound formed when DDT breaks down in the environment, prevents normal calcium deposition when eggshells are forming. The thin-shelled eggs are then susceptible to breakage during the incubation period. Few chicks survived in this era. Bald eagles and other raptors were vulnerable to DDT/DDE poisoning because toxic substances bioaccumulate, or build up, in the food chain and these predators sit atop that chain.

Thousands of bald eagles were also shot during the 20th

century, many in the belief that they would steal livestock. Although I have seen a bald eagle flying down the road on Pender Island with a chicken clutched in its talons, this kind of livestock predation is rare. Eagles are almost always scavengers. In 1782, the bald eagle became America's national symbol over the objections of Benjamin Franklin, who preferred wild turkeys. Franklin felt eagles' propensity for eating carrion reflected "bad moral character." By the early 1960s, there were as few as 400 pairs of bald eagles nesting in the lower 48 states. Many states, from Nevada to New Hampshire, had no eagles at all.

The Migratory Birds Treaty signed by Canada and the U.S. in 1916 offered a modicum of protection from hunting, supplemented by 1940's American Bald Eagle Protection Act, which prohibited killing or possessing eagles for commercial purposes. Bald eagles were formally declared endangered in the U.S. in 1967. Laws banning the use of DDT were enacted in the 1970s. Bald eagle chicks began to survive, and populations began to recover. The U.S. Fish and Wildlife Service down-listed bald eagles from endangered to threatened in 1995, and by 2007 they no longer required the protection of the Endangered Species Act.

In January 1994, I visited the annual eagle festival held at Brackendale, a small community about an hour's drive north of Vancouver. With salmon running in the Squamish River, bald eagles congregate to feast on the rotting carcasses of spawned-out chum and coho. Cottonwood trees along the river were leafless because it was the middle of winter, but appeared decorated for Christmas with strings of white beads. These were the bright white heads of bald eagles perched in their branches, taking a break from the salmon buffet. Expert birders conducting the annual eagle census at Brackendale on January 9, 1994, counted 3,769 eagles, a world record that stands to this day. In the course of a brief afternoon hike along the riverbank, I saw more bald eagles than had lived in the entire continental United States 25 years earlier. There are now more than 100,000 bald eagles in the U.S., nesting in every state except Hawaii.

ANOTHER BIRD WHOSE EXISTENCE was jeopardized by DDT is the peregrine falcon, a raptor exquisitely evolved for aerial hunting. The size of a crow, with a three-foot wingspan, these falcons have a sickle-shaped silhouette and a notched beak used to sever the spinal column of their prey. Peregrines are the fastest animals in the world. Olympic gold medallist Usain Bolt of Jamaica can sprint 43 kilometres per hour. Triple crown–winning racehorse Secretariat thundered down the home stretch at 79 kilometres per hour. A cheetah in hot pursuit of a gazelle can briefly reach 97 kilometres per hour. None are even in the same league as the peregrine, whose predatory dives can reach speeds of over 435 kilometres per hour. Peregrines have an extra eyelid to protect their eyes, spread tears, and clear away debris while maintaining their vision at such extreme speeds. Specially adapted nostrils enable peregrines to breathe despite the extreme air pressure created by high-speed dives.

The peregrine's blazing speed was no defence against decades of DDT use, which eliminated them from vast swaths of North America. None were left in the southern half of Canada and only a few pairs remained in remote regions of the Yukon and Northwest Territories. In the U.S., peregrine populations declined by more than 90 percent.

Although naturally much less common than bald eagles,

peregrine populations have also recovered. Canada and the U.S. began captive breeding programs in the 1970s, releasing peregrines back into the wild. Biologists used various strategies in the recovery effort, from artificial insemination to hacking (training young falcons to hunt) and rappelling down cliffs to place captive-bred chicks in active falcon nests. The first active nest of a new generation of peregrines in Ontario was located in 1986, and every year since then, their numbers have increased. There are now thousands of nesting pairs in Canada and the U.S. In 1999, peregrine falcons were delisted from the U.S. Endangered Species Act. In 2007, the Committee on the Status of Endangered Wildlife in Canada down-listed peregrines to a species of special concern.

WHALES AND RAPTORS TEACH US two simple conservation lessons. Stop hunting them and stop poisoning their environments, and they will probably recover. A more complex challenge is to ensure that people share land, water, and ecosystems that provide habitat critical to the survival and recovery of species at risk. Human alteration and destruction of habitat is the main reason why species become endangered today. Converting prairies to agricultural land, logging old-growth forests, and building cities in rich coastal zones have taken a toll. But here too there are encouraging signs. In North America and around the world, there has been exponential growth in the volume of land set aside in parks and protected areas, places where industrial activities, farming, ranching, and urban development are by and large prohibited.

One of the first campaigns I worked on as a young environmental lawyer was an effort to protect B.C.'s Tatshenshini River from a proposed open-pit copper mine at Windy Craggy

Mountain. The Tat is located in the farthest reaches of northwest B.C., next to the Yukon's Kluane National Park and Alaska's Glacier Bay and Wrangell–St. Elias national parks. I'd resolved to try to visit as many of the places that I worked to save as possible, so I joined a 10-day wilderness rafting expedition down the river. Those rafting trips were a key part of the campaign, because anyone who saw that place could not help but fight to protect it. The Tatshenshini watershed is spectacular, a masterpiece of craggy, mountainous beauty that achieves the impossible by ratcheting up the awesomeness level every day on the river, culminating in the iceberg-strewn Alsek Lake. It has high ecological values as a healthy wild salmon river frequented by grizzly bears, wolves, and other predators. B.C. Premier Mike Harcourt and his government eventually made the right decision, creating a massive new park spanning nearly one million hectares. The Tat is now the jewel in the crown of a World Heritage Site that, combined with neighbouring parks, forms the largest protected area on Earth.

During the 1990s, British Columbia doubled the protected area of the province from 6 percent to 12 percent, creating hundreds of dazzling new parks that safeguard vital wildlife habitat. During the same decade, governments in Canada designated almost one hundred million acres of new parkland, bringing our protected area percentage up to almost 10 percent. Where I live, the federal government established the Gulf Islands National Park, encompassing parts of Pender and Saturna as well as a number of smaller islands. There are endangered ecosystems here, such as Garry oak meadows, once common on south facing slopes, with gorgeous assemblages of Taylor's checkerspot and swallowtail butterflies, and native flowers like chocolate lilies and blue camas. A whopping 43 species native to this ecosystem are listed under Canada's Species at Risk Act. These meadows have been overrun by invasive species like Scotch broom, which covered our land in a thick blanket when we first moved to Pender. It took many hours of gruelling labour to pull the broom plants from the ground, roots and all. The broom left a legacy of thousands of seeds, providing us with an annual opportunity for sore hands and stiff backs. We planted a Garry oak seedling 14 years ago, with a wire cage around it to stave off white-tailed deer. This year, the slow-growing tree reached my knee.

As impressive as B.C.'s and Canada's recent accomplishments in setting aside land may seem, they pale in comparison to the leadership shown by other countries. More than 70 nations protect a larger percentage of their land base than Canada. The leading coutries, each of which has set aside at least one-quarter of their land base, include Brazil, Bhutan, Costa Rica, the Czech Republic, New Zealand, the Seychelles, Slovenia, Spain, and Switzerland. The establishment of tens of thousands of new parks—in ecosystems ranging from the mountains of Nepal to the rainforests of Brazil—is an immensely encouraging development. Even in China, where the first national park was not established until 1982, more than a century after Yellowstone and Banff, there are now almost 2,000 national parks protecting millions of hectares from industrial development. In total, over 15 percent of the world's land—more than 25 million square kilometres (the size of North America)—is now protected, a dramatic development that occurred mainly during the past century.

Protecting marine habitat is equally important. Canada has made repeated promises to establish a network of marine protected areas (MPAs) that rivals our national parks, but progress has been painfully slow, with less than 1 percent of Canada's huge marine area protected. At the other end of the spectrum, Monaco, a tiny country with an equally small coastline, leads the world with 100 percent of its maritime territory protected. Portugal has set aside over half of its ocean realm in MPAs. Australia, with a large ocean territory like Canada, has designated 35 percent of its oceans for protection. The tiny nation of Kiribati, located halfway between Fiji and Hawaii in the South Pacific, established an MPA the size of California around the Phoenix Islands. According to *National Geographic*'s explorer-in-residence Enric Sala, diving in Kiribati's Phoenix Islands MPA "is like getting in a time machine and travelling back to the reefs of the past, when sharks—and not humans—were the top predators."

In the U.S., recent presidents have played a game of MPA one-upmanship. President Clinton issued an executive order in 2001, calling for the creation of a national system of MPAs. President Bush, who was otherwise pretty dismal from an environmental perspective, set aside the 363,000 square kilometres Papahānaumokuākea Marine National Monument in Hawaii in 2006 and three smaller reserves in 2009. Papahānaumokuākea is home to more than 7,000 species ranging from green turtles to albatrosses, and is designated as a World Heritage Site. In 2014, President Obama, burnishing his green reputation, designated the world's largest MPA, the Pacific Remote Islands Marine National Monument. Covering two million square kilometres far from the nearest human settlement (eight days by ship from the nearest port), this reserve is off limits to commercial fishing, drilling for fossil fuels, or mineral exploration. Some of its corals are up to 5,000 years old, millions of seabirds nest on the islands and atolls, and the abundance of fish is reputedly incredible (including the quirky bumphead parrotfish that engages in ritualized head-butting similar to bighorn sheep).

Conservation biologists estimate that approximately half of most ecosystems need to be protected in order for the full diversity of life on Earth to continue flourishing and evolving. This doesn't necessarily require that lands and waters be off limits to all forms of human activity. Often there are creative solutions. My good friend Scott Wallace, a marine biologist who works for the David Suzuki Foundation, led the negotiation of a precedent-setting agreement with B.C.'s groundfish bottom trawling fleet that includes strong incentives to limit fishing in areas that are important habitat for sensitive corals, sponges, and other essential elements of marine ecosystems. A tight bycatch quota of 20 kilograms of combined corals and sponges per trawl was agreed upon, the most conservative in the world. The deep-sea area open for trawling is 65 percent smaller than in the past. Every boat carries an independent observer to monitor its catch. Nature is protected, the fishermen are happy, and it seems like a win-win situation.

WHILE WE ARE MAKING PROGRESS in protecting habitat for endangered species (and to prevent wildlife from becoming endangered), a more daunting and expensive task is to restore damaged ecosystems and reintroduce extirpated species. For all of our technological wizardry, humans cannot hold a candle to four billion years of evolutionary experimentation and adaptation. We can create artificial lakes and wetlands, but they're often a shadow of Nature's originals. Nevertheless, we must try.

One of Canada's most challenging endangered species recovery efforts involves a small patch of desert in B.C.'s South Okanagan.

This is the northernmost tip of the Sonoran Desert, which stretches all the way to Mexico. It is not a desert of stately saguaro cactuses but a more modest ecosystem where prickly pear cacti provide a spiked ankle-height carpet stretching across the valley. Much of the desert has been converted to agriculture, primarily fruit orchards and vineyards. Margot and I went on a tour of the desert east of Osoyoos with a guide from the Okanagan First Nation. It was a sweltering summer day, with temperatures pushing 40 degrees Celsius and not even the hint of a breeze. He recommended that we carry three items—water, an umbrella (DIY shade in a place with no trees), and a fork. The fork, it turned out, was handy for removing cacti from your calves. The natural motion of walking on these plants causes your heel to flick prickly pear cactus projectiles upwards, where their needle-sharp spikes would embed in the back of your leg. You couldn't use your hand to pull them off without additional self-inflicted acupuncture, but the fork worked like a charm. As luck would have it, the desert was in bloom the day of our visit: tiny orange-yellow blossoms were like a constellation of stars on the ground, attracting hordes of hummingbirds to an ephemeral feast. The goal for restoring this ecosystem is to protect and reconnect a few pockets of desert, creating a continuous zone for the area's wild creatures. The long-term hope is to rebuild populations of endangered native species, including badgers and burrowing owls. It is an ambitious project.

Prairie ecosystems also epitomize the magnitude of the challenge in restoring wildlife habitat lost to human habitation and use. Three types of grasslands (tall grass, short grass, and mixed grass) once covered a vast swath of North America's interior, but only tiny fragments escaped the settler's plow. To remedy this, land purchases and conservation agreements with farmers and ranchers are restoring wild native grasslands. This critical habitat supports the yeoman efforts being made to protect endangered species like Greater Prairie chickens, whose mating dance would be a runaway winner on *America's Got Talent* if wild species were eligible. Where millions of bison once wandered free, now there are fences everywhere. Yet bison, swift foxes, burrowing owls, and black-footed ferrets are being reintroduced on both sides of the Canada-U.S. border and unfenced areas are being encouraged. The wood bison was down-listed from threatened to special concern in 2013 because of its growing population. Thousands of plains bison live in large parks, with Parks Canada planning to bring them back to Banff National Park.

Like the bison, millions of black-footed ferrets once prowled the grasslands, from southern Canada to Texas. These ferrets have a white face with a black mask, and depend on prairie dogs for food and shelter (taking over the latter's underground burrows). They were nearly driven extinct by the shooting, poisoning, and deliberate elimination of prairie dog colonies by farmers and ranchers. Feared to have vanished forever in the 1960s, a small population was found in South Dakota and used to start a captive breeding program. By 1979 that program had failed, and once again the black-footed ferret teetered on the brink of extinction. In 1981, a tiny, relic population was discovered in a prairie dog colony near Meeteetse, Wyoming. Government agencies, with the help of landowners, NGOs, and Indigenous people, collected all remaining wild ferrets and started a second captive breeding program. This time it was successful, and in the U.S. there are now 1,500 ferrets living in the wild. Parks Canada released ferrets in Grasslands National Park in 2009 and the Canadian reintroduction is ongoing.

CYNICS MAY SCOFF that I have cherry-picked a few species of charismatic megafauna that have benefitted from extensive and expensive recovery efforts. Not true. Dozens of species have been removed from lists of endangered wildlife in Canada and the U.S. More than 30 species have been de-listed in the U.S. including the bald eagle, peregrine falcon, gray whale, grizzly bear, gray wolf, brown pelican, Stellar sea lion, American alligator, a snake, a flycatcher, a flying squirrel, a lizard, a fish, an orchid, and a daisy. The Center for Biological Diversity identified more than 20 species whose populations increased by more than 1,000 percent in recent decades. There was a 2,206 percent increase in nesting Atlantic green sea turtle females on Florida beaches. The California least tern enjoyed a 2,819 percent increase in nesting pairs. The San Miguel island fox population increased 3,830 percent. Numbers of the El Segundo blue butterfly increased 22,312 percent. Studies indicate that roughly 90 percent of species listed under the U.S. Endangered Species Act are on track to meet their recovery targets by the projected deadline. Imagine other government programs—reducing poverty, rehabilitating criminals, subsidizing businesses, etc.—achieving a 90 percent success rate.

In Canada, creatures no longer requiring the protection of the Species at Risk Act include the white pelican, Baird's sparrow, and Caspian tern. Dozens more have been down-listed to less dire categories to reflect their improving circumstances, including sea otters, humpback whales, wood bison, and the Rocky Mountain tailed frog. In the Salish Sea, harbour seals have rebounded from a brush with extirpation and are now flourishing, while Pacific white-sided dolphins have returned in force after a century-long absence. These heart-warming advances resulted from hard work by biologists, activists, government officials, businesses, and property owners.

The good news is not limited to Canada and the U.S. Using trends in the size of more than 9,000 populations of 2,688 mammal, bird, reptile, amphibian, and fish species from different biomes and regions, the World Wildlife Fund determined that there has been a 31 percent increase since 1970 in temperate regions of the world. That is not a typo—the WWF determined that across a broad range of native species in North America and Europe, wildlife populations have increased substantially over the past 40 years. That part of the WWF study got little media attention. Instead, the headlines concentrated on a 60 percent decline in wildlife populations in tropical regions over the same time frame. Of course this bad news is deeply troubling, reflecting the impact of population growth, economic growth, and the North-South shift in environmental pressure to support overconsumption in wealthy countries. We cannot solve environmental problems by transferring the consequences of our excessive consumption to developing countries. Genuine solutions will require substantial reductions in resource use in the North, through behavioural changes and technological innovations that do not cause harm in the South.

Much of North America and Europe is undergoing reforestation, with forests being replanted or regenerating naturally. Programs and policies to reduce tropical deforestation, and avoid the associated global warming emissions, are bearing fruit in 17 countries across four continents, according to a recent study by the Union of Concerned Scientists. The countries studied include Guyana, Brazil, Kenya, Madagascar, Costa Rica, Mexico, and Vietnam. In 2011, the Philippines enacted a total ban on logging in natural forests, permitting trees to be cut only on plantations. This action was motivated by concerns about climate change, landslides, and a desire to ensure that future generations of Filipinos would be able to enjoy old-growth tropical forests and their rich biological diversity. Another impetus for the Philippines' leadership was a 1993 decision from their Supreme Court in a groundbreaking lawsuit filed on behalf of children and future generations by lawyer Tony Oposa, Jr. The case asserted that clear-cutting old-growth forests violated the constitutional right to live in a healthy environment, and the court's powerful judgment led to the cancellation of many logging contracts.

A recent article in *Science* described how between 2004 and 2014, farmers and ranchers in the Brazilian rainforest have saved tens of thousands of square kilometres from clear-cutting (the equivalent of almost 14 million soccer fields). Overall, Brazil has used its powerful green constitution, strong laws, and improved enforcement to reduce deforestation by 70 percent. Because tropical rainforests inhale carbon dioxide and convert it to oxygen, this decrease in logging has kept more than three billion tonnes of greenhouse gases from entering the atmosphere, akin to taking every car in the U.S. off the roads for a year.

Grassroots reforestation efforts in China have created what former Chinese leader Deng Xiaoping called the Great Green Wall, a band of forests intended to block the spread of deserts. Over 26 million hectares have been successfully reforested since the project was launched in 1978. One of the leaders of the movement is Shi Guangyin, a peasant whose life was altered by an experience he endured as an eight-year-old boy. He was herding sheep near his home in Dingbian County in China's Shaanxi Province when he was caught in a terrible sandstorm. Shi narrowly escaped death but his friend, five-year-old Zhao Huwa, perished in the sandstorm. Now an elderly man, he still remembers the nightmarish event, saying, "I knew I had to push back the sand to survive."

In Kenya, the Greenbelt movement for which Wangari Maathai won a Nobel Peace Prize in 2004 has planted over 51 million trees. The grassroots organization sponsors 4,000 tree nurseries that produce over eight million native seedlings annually. More than 30,000 women have received training in forestry, bee-keeping, food processing, and other trades that enable them to earn a livelihood while protecting local lands and ecosystems. Similar movements now exist in Tanzania, Uganda, and other African nations. Maathai's advocacy also inspired the UN to undertake its billion tree campaign, which has led to the planting of over twelve billion trees since its launch in 2006.

A SMALL MINORITY OF SCIENTISTS believes that the planet is so large and its ecosystems so resilient that human actions cannot possibly cause lasting ecological harm. For example, Peter Kareiva, a scientist who used to work for the Nature Conservancy, argued, "Nature is so resilient that it can recover rapidly from even the most powerful human disturbances." I completely disagree with such a blanket statement, which would give us the green light to ignore the ecological consequences of our actions. Some species are incredibly resilient: think cockroaches, weeds, rats, and humans. Others are relatively vulnerable: rare orchids, large mammals, coral reefs, and flightless birds.

Sometimes a species will surprise us, offering a valuable reminder of how much we have yet to learn, or may never know, about the natural world. Consider the Quino checkerspot butterfly, which exhibits a colourful checkerboard of yellow, orange, red, and brown spots and is found in Mexico and California. Because of climate change, experts feared that the Quino would no longer be able to find the single plant upon which it laid eggs and would thus die out. Professor Camille Parmesan of Plymouth University said, "Every butterfly biologist who knew anything about the Quino in the mid-1990s thought it would be extinct by now, including me." Another biologist described its status as an airplane with "four engines out and ten seconds to impact." Confounding the expectations of scientists, this butterfly appears to have shifted its range to higher altitudes and selected a new plant upon which to lay its eggs. It has, so far, successfully adapted to climate change. Parmesan added, "In the early days of climate change, people worried that nature reserves would no longer be useful because the species they protected would move out. Now we know that new species move in, and so they are more important than ever."

HUMANS CAN ASSIST THESE incredible species recoveries, though we're not out of the woods by any means. There are still thousands of species on endangered and threatened lists around the world, and sometimes our recovery actions are too little, too late. We've been an awful neighbour, but we've apologized, matured, and promised to do better. Millions of people, living in communities in every country, are actively engaged in promoting habitat restoration and the recovery of endangered species. Perhaps humans are finally recognizing that we are part of, not separate from, the community of life on Earth, and are embracing the responsibilities that accompany our technological prowess.

*"If I keep on saying to myself that I cannot do a certain thing,
it is possible that I may end by really becoming incapable of doing it.
On the contrary, if I have the belief that I can do it, I shall surely acquire
the capacity to do it even if I may not have it at the beginning."*
—Mahatma Gandhi

Chapter 2
The Renewable Energy Revolution

OIL, NATURAL GAS, AND COAL are largely responsible for the advanced state of industrial society and much of our wealth. The energy density of these fossil fuels makes them tremendously useful: a single litre of gasoline contains the energy equivalent of 10 to 20 days of human labour. Oil dominates the global transportation system. Coal is still the single largest source of electricity, generating more than 40 percent of the world's supply. Weaning society from our fossil fuel addiction is a process that will take decades, yet the transition to clean, renewable energy is inevitable and accelerating. We've already passed the peak of conventional oil, with a growing proportion of today's oil coming from costly, harder to reach places—tar sands, shale, and deep offshore deposits. And don't forget the well-known health and ecological damage inflicted by burning fossil fuels and the potentially catastrophic effects of climate change. The real question is: How fast can we make the tectonic shift from burning fossil fuels to harvesting clean energy?

Although coal is still the number-one source of electricity worldwide, the coal era is ending. In Canada, Ontario shut down all of its coal-fired electricity plants, replacing them with a mix of solar, wind, and natural gas. Other provinces are phasing out coal plants. The federal government prohibited the construction of new coal plants unless their emissions are as low as natural gas plants, which is not even close to being economically feasible at this time. In the U.S. between 2010 and 2015, 175 coal-fired electricity plants either shut down or announced they would be closing soon. Almost 200 proposed coal power plants in the U.S. have been cancelled. New regulations targeting emissions of mercury and carbon dioxide will force the closure of additional American coal plants. The demise of coal will be an enormous victory for cleaner air, a stable climate, human health, and ecosystem integrity.

The renewable energy revolution is happening faster than anyone predicted and faster than most people grasp. Even experts are surprised by the speed of the shift. The change is accelerating because of concerns about climate change, peak oil, and health impacts, and also thanks to technological advances and the plummeting costs of renewable energy. The International Energy Agency (IEA), whose 29 members comprise most nations of the Organization for Economic Cooperation and Development (OECD), produces a report called *World Energy Outlook*, described as "the authoritative source for projections of the world's energy future." Despite the blue-ribbon nature of the IEA, their forecasts have consistently and dramatically underestimated the ascent of renewables. In 2000, the IEA predicted that global wind power would reach a total of 30 gigawatts by 2010. (A gigawatt is equal to a billion watts or 1,000 megawatts. Large power plants are often in the 500–1,000 megawatt range.) Oops! Their forecast was exceeded in 2003, seven years early, and more than 30 gigawatts of wind power has been installed *every year* since 2009. The IEA predicted 63 gigawatts of wind globally by 2020, a total surpassed by Europe alone back in 2008. The IEA's forecast for 3.7 gigawatts of Chinese wind power by 2020 has already been exceeded 30 times over, with China adding 23 gigawatts in 2014 alone. The same absurd conservatism plagues the IEA's solar energy projections. There was one gigawatt of installed solar capacity in the world as of 2000. The IEA estimated that global solar plus tidal plus wave plus other emerging renewables would be 7.6 gigawatts by 2020. By the end of 2014, the global total for solar alone was 150 gigawatts. In other words, in just over a decade, the amount of electricity generated by solar panels increased 150 times. Globally, installed solar capacity has been rising 50 percent per year since 2006, slowing down only slightly during the recession. China alone is adding more sun-powered electricity *every six months* than the IEA's predicted 2020 total for the whole world.

We shouldn't be too hard on the IEA. People generally have difficulty grasping the meaning and implications of exponential growth. Exponential growth means that a quantity grows by multiplying, whereas linear or arithmetic growth (which we do understand) increases by addition. Malcolm Gladwell illustrates the concept well in his book *The Tipping Point*:

> Consider, for example, the following puzzle. I give you a large piece of paper, and I ask you to fold it over once, and then take that folded paper and fold it over again, and

then again, and again, until you have refolded the original paper 50 times. How tall do you think the final stack is going to be? In answer to that question, most people will fold the sheet in their mind's eye, and guess that the pile would be as thick as a phone book or, if they're really courageous, they'll say that it would be as tall as a refrigerator. But the real answer is that the height of the stack would approximate the distance to the sun.

Gladwell's example is hard to believe, and I spent an hour with a calculator to overcome my own skepticism. While the world is still in the early stages of a massive energy switch, renewable energy is growing at an exponential rate. The implications of exponential growth for solar-powered electricity are almost deliriously exciting. Even if the growth rate slows to 35 percent annually, the total amount of installed solar capacity will double every two to three years. By 2050, today's 150 gigawatts of solar could increase to over 1,000,000 gigawatts, easily enough to meet the world's electricity needs many times over. As Thomas Edison once sagely observed, "I'd put my money on the sun and solar energy."

Cellular telephones illustrate the potentially disruptive impact of new technologies experiencing exponential growth. I remember laughing out loud the first time I saw someone walking down the sidewalk, speaking into a brick-sized device with a big antenna sticking out of it. The gap between the first mobile phones, which weighed more than two kilograms, and today's nifty little smartphones is enormous. In 1980, telecommunications giant AT&T enthusiastically predicted that there would be a market of approximately 900,000 subscribers for mobile phones by the year 2000.

Not even close! There were over 100 million. Fast forward to 2015, and there are over six billion cell phone contracts in the world.

SOLAR PHOTOVOLTAIC (PV) SYSTEMS, which convert sunlight into electricity, have ridden a rollercoaster over the past four decades. Solar seemed like a slam dunk in the early 1970s because of the OPEC oil embargo, the spike in energy prices, and the emergence of the environmental movement. During the second OPEC oil crisis in 1979, President Jimmy Carter put a 32-panel solar hot water system on the roof of the White House. Carter predicted that "a generation from now, this solar heater can either be a curiosity, a museum piece, an example of a road not taken, or it can be a small part of one of the greatest and most exciting adventures ever undertaken by the American people; harnessing the power of the sun to enrich our lives as we move away from our crippling dependence on foreign oil." Then energy prices crashed. Ronald Reagan replaced Carter as president. Reagan and his chief of staff, Donald Regan, felt the solar equipment was a joke, and it was removed. Since then, oil prices have skyrocketed and gyrated, climate change has emerged as a global peril, and solar costs have dropped like a stone. President Barack Obama recently reinstalled a solar hot water heater on the White House and added a 6 kilowatt solar PV system. Although this chapter focuses on renewable electricity, it's worth noting that solar hot water systems, which use the sun's energy to heat water for domestic use, are cost-effective and can be used in almost any climate. Building codes in many areas—at the national level in Spain, Portugal, Israel, and municipally in parts of Germany, Ireland, Italy, the U.S., and Canada—make solar hot water systems mandatory.

Our ability to convert sunshine into electricity has become much cheaper more rapidly than anyone predicted. Solar's overall share of the global energy pie is still a very small slice, but it is growing rapidly. When David Suzuki and I co-authored his *Green Guide* in 2008, the largest solar photovoltaic power plant in the world was a German facility with a capacity of 40 megawatts. Within just six years, there were more than 80 solar PV plants larger than 40 megawatts (in Europe, India, China, Canada, South Africa, and the U.S.), with the largest being 550 megawatts (the Topaz Solar Farm in California). Several solar PV facilities under construction or in the planning stages are in the 2,000 to 10,000 megawatt range, which would place them among the world's largest electricity-producing facilities.

Concentrating solar power systems employ a different technology, using mirrors to reflect sunlight from a large area onto a central tower. The light is converted into heat, producing steam that turns turbines and generates electricity. While the world's largest concentrating solar power plant in 2008 was 80 megawatts, now there are several larger than 350 megawatts. The Ivanpah Solar Electric Generating System in southeast California is a 392 megawatt power plant that began operating in 2014. Viewed from the sky, Ivanpah looks like a massive modern art installation. Three large circles, each comprised of rings of thousands of mirrors mounted on the ground to reflect sunlight, surround three iconic towers. It's a beautiful sight, particularly in comparison to a coal-burning power plant. At a construction cost of $2.2 billion, the facility is expected to generate enough electricity to power 140,000 homes. Four other utility-sized concentrating solar power plants came online in the southwestern U.S. in 2014, marking an unprecedented leap forward for this new technology. There are some concerns about bird deaths caused by these facilities, but the reality is that domestic cats kill many orders of magnitude more birds.

One of the common criticisms levelled against solar power is that people and businesses like to continue using electricity after the sun goes down. Yet, progress is being made in overcoming this challenge. In 2011, the Gemasolar concentrating solar power plant located on the plains outside the Spanish city of Seville became the first in the world to generate electricity 24 hours a day, seven days a week. Gemasolar broke new ground by diverting some of the heat produced during the day into huge insulated vats of molten salt, then converting the heat into electricity during the night. Concentrating solar power with storage is also being used at the Solana project in Arizona and the Crescent Dunes power plant in Nevada. Many options for storing solar-generated electricity are being developed, from improved batteries to flywheels.

The Drake Landing Solar Community in Okotoks, Alberta, stores the sun's warmth underground during the summer and provides heat for 52 homes during the winter. An array of 800 panels captures solar energy and transfers the heat to a borehole thermal energy storage system, where water in underground pipes warms the surrounding earth. By the end of summer, the underground temperature reaches 80 degrees Celsius. To maintain this heat, the system is covered with sand, high-density insulation, a waterproof cover, clay, and landscaping materials. During winter, the hot water is circulated to homes through a district heating system. While this kind of system was pioneered in northern Europe, the Drake Landing project is the first in the world to provide over 90 percent of the homes' space heating requirements through stored solar energy. Given that average temperatures in Okotoks are below freezing for five months a year, this is an impressive accomplishment!

People also wonder if there's enough solar energy to power humanity. The answer is an emphatic yes. Scientists calculate that the amount of solar energy reaching the Earth's surface every hour exceeds global energy consumption for the entire year. Of course we're not about to cover the whole planet with solar panels, but you get the idea. A proposed series of solar electricity plants located in the Sahara Desert could in theory provide enough electricity for the entire continents of Africa and Europe. An international consortium is exploring the feasibility of this concept.

Our ability to emulate plants by turning sunshine into useable energy has improved dramatically—and likely will continue to do so. From 2008 to 2015, the price of solar panels dropped 80 percent—a faster decline than even the most bullish prognosticators had anticipated—thanks to economies of scale from increased levels of production (especially in China), experience, and technological advances. Installation costs are also declining as workers gain experience, as illustrated by significantly lower costs in Germany compared to the U.S. and Canada.

How disruptive is the potential power of solar energy? According to the *New York Times*, "Electric utility executives all over the world are watching nervously as technologies they once dismissed as irrelevant begin to threaten their long-established business plans." The rapid decline in the costs of solar-powered electricity is creating a "utility death spiral" because the business model that generated tremendous profits for electric utility shareholders for more than a century is in mortal danger. Each time another solar PV system is installed on a home or business, the market share of companies that produce and distribute electricity from large, centralized power stations shrinks a little bit further. The electric utility's costs are spread across a diminishing customer base, their credit worthiness declines, financing costs increase, and prices rise, sending even more people into the waiting arms of the burgeoning solar industry. More solar PV systems are installed and the process repeats itself.

The top 20 utility companies in Europe have lost more than half of their value since 2008. Some of this decline can be blamed on the recession, but other sectors have recovered, while big utility companies have not. In 2013, Germany's largest power company, RWE, lost money for the first time since 1949. After announcing a $3.8 billion loss, RWE's chief executive Peter Terium admitted, "We were late entering into the renewables market—possibly too late." In Australia, both public and private utility companies are struggling to make profits and are laying the blame on solar. George Bilicic, global head of power, energy, and infrastructure for the international investment bank Lazard said, "We used to say some day wind and solar would be competitive with conventional generation. Well, now it is some day."

In 2013, the Edison Electric Institute, a lobby group representing American utility companies, warned the industry that it had waited too long to respond to the sharp cost declines and growing popularity of solar: "At the point when utility investors become focused on these new risks and start to witness significant customer- and earnings-erosion trends, they will respond to these challenges. But, by then, it may be too late to repair the utility business model." In 2014, Barclays bank downgraded its ratings for the entire U.S. utility sector because of "increased recognition of the long-term threat to grid power." What did Barclays mean by long-term? In the bank's own words, "We believe that solar plus storage could reconfigure the organization and regulation of the electric power business over the coming decade." Not the coming century, the coming decade.

The Holy Grail for solar is something called grid parity, when the cost of producing electricity from sunlight declines to the point where it's on par with the cost of electricity generated by coal, natural gas, or nuclear energy. By on par, I mean in the narrow economic sense of dollars spent to produce the electricity, not including the many negative externalities such as carbon pollution, air pollution, deaths, illnesses, military spending to ensure access to oil, or the cost of safely managing nuclear wastes that are not incorporated in conventional electricity prices. The Holy Grail moment for solar has already arrived in a growing number of jurisdictions, particularly those with sunny climates and high electricity prices, such as Australia, California, Hawaii, and broad swaths of Africa, South Asia, Latin America, and the Middle East. Experts have predicted that by 2020, 80 percent of the world's population will live in regions where prices for solar electricity are no higher than electricity from more environmentally destructive sources. Solar has always been a green, future-friendly choice. But today's explosion in solar power is being driven by economics, not altruism. Consumers are shifting to solar electricity because it's cheaper or costs the same.

Accelerating this transformation are financial businesses that are creating decentralized utilities one rooftop at a time. Among the options are solar leases, loans, and power purchase agreements.

For example, SolarCity, founded by Elon Musk, offers to install PV systems on residential and commercial buildings at no cost to the building's owners. The owners simply sign a long-term electricity purchase contract with locked-in rates that are guaranteed to be less than the price charged by the utility company serving the building. Similarly, the second largest homebuilder in the U.S., Lennar Corporation, is now putting solar PV systems on the roof of every house it constructs in California, Colorado, and several other states. Lennar installs the systems, retains ownership of the panels, and sells the electricity to the homebuyer at a price guaranteed to be 20 percent below the local utility's price for the next 20 years. As David J. Kaiserman, president of Lennar's solar division, said, "It's so simple when we tell a customer, 'You're guaranteed to save money.'"

LARGE-SCALE WIND POWER is even cheaper than utility-scale solar, especially land-based wind farms, which are much less expensive than offshore facilities. Like solar, wind is enjoying exponential growth. The global installed wind capacity at the end of 2014 was 370 gigawatts, more than 70 times higher than it was just two decades ago. That figure is expected to grow to 2,000 gigawatts by 2030, enabling wind to supply almost 20 percent of the world's electricity needs. According to Steve Sawyer, CEO of the Global Wind Energy Council, "Wind power has become the least-cost option when adding new capacity to the grid in an increasing number of markets, and prices continue to fall." China generates more than one-quarter of the world's wind energy, followed by the U.S., Germany, Spain, and India. China installs a new wind turbine every hour. The U.K. and Denmark are leaders in offshore wind production, despite the higher costs. Globally, offshore wind has grown from less than 1 gigawatt in 2006 to 8 gigawatts today and will increase to an estimated 40 gigawatts by 2020. The Alta Wind Energy Center in the Tehachapi Mountains of California is currently the world's largest wind farm, with an installed capacity of 1,320 megawatts and a planned capacity of 3,000 megawatts.

Of course, as is the case with all forms of energy, wind has its detractors. Even in the most blustery locales, the wind does not always blow, meaning wind must be combined with another source of power. Wind dovetails nicely with hydroelectricity, as reservoirs can serve as de facto batteries. There are collisions between turbine blades and some species of birds and bats. Careful siting decisions and improved technology have alleviated, but not eliminated, these problems. Domestic cats and office buildings kill far more birds than wind turbines. Some people report adverse health effects from living in close proximity to wind farms, yet airports and major roads expose people to more noise pollution than wind turbines. All told, the health and environmental effects of producing energy from wind are modest when compared with the immense costs of burning fossil fuels, the risk of catastrophic climate change, or the dangers of relying on nuclear energy. Although some high-profile environmentalists—such as James Lovelock—support nuclear energy, it faces daunting obstacles, including the risk of disastrous accidents or terrorist attacks, the unsolved dilemma of storing nuclear waste, and very high costs.

It is important to emphasize that in wealthy industrialized countries, adding renewable energy capacity should always be complemented by major efforts to reduce overall energy use. Even green energy inflicts environmental damage. Public policies, including strong regulations and pollution taxes, must be used to ensure greater energy efficiency and to spur behavioural change in key areas such as transportation choices and housing size. Europe is far ahead in using well-designed taxes to drive the shift to more efficient vehicles, as well as more trips taken on foot, by bike, and on public transit. North American cities need to encourage increased population density while discouraging sprawl. Reducing energy use will make it easier to generate 100 percent of the electricity supply from renewable sources and minimize the negative environmental impacts inevitably caused by clean and green energy.

LOOKING AT THE RENEWABLE ENERGY REVOLUTION on a country-by-country basis, Canada has the potential to meet 100 percent of its energy needs, thanks to tremendous wind, solar, geothermal, biomass, wave, and tidal energy resources. According to a report co-authored by the Canadian Academy of Engineering and the David Suzuki Foundation, these resources "are many times larger than current or projected levels of total fuel and electricity consumption." As of 2014, almost two-thirds of Canadian electricity consumption was generated from renewable sources. Sixty percent of Canada's electricity was produced by hydroelectricity, which is widely regarded as clean energy but has substantial environmental impacts and caused the forced relocation of Aboriginal communities. Wind, biomass, and solar photovoltaic trail far behind at 1.6 percent, 1.4 percent, and 0.5 percent respectively. Burning fossil fuels generates roughly 25 percent of Canadian electricity, and nuclear provides the remaining 15 percent. Transportation is almost entirely fossil fuel based, although electric vehicle sales are growing rapidly.

I remember seeing my first wind farm, in southwestern Alberta, in 1993 while en route to a backpacking trip in the Rockies. The Cowley Ridge wind project was the first in Canada, with about 50 turbines. It's a constantly windy landscape, where people who do not share my fear of heights charter planes to tow their gliders into the air and catch updrafts to insane heights of 40,000 feet. (Yes, they're more than 10,000 feet above Mt. Everest in a lightweight aircraft with no engine.) I stopped along the highway to have a closer look. The tall turbines gleamed in the morning sun, their blades spinning steadily in the breeze blowing down off the mountains. Cattle grazed in their long shadows. There was a low hum from the blades but it was swept away by the steady thrum of the wind itself.

Wind and solar have experienced blistering growth in Canada in recent years, while wave and tidal power are just beginning to be deployed. In 1997, Canada had 23 megawatts of installed wind capacity. By 2015, that number had grown to almost 9,700 megawatts, placing Canada seventh in the world. In 2014, Canada ranked sixth in adding new wind capacity, behind China, Germany, the U.S., Brazil, and India. Thirty-seven new wind projects were completed in 2014. The International Energy Agency expects Canada's total will grow by another 50 percent in the next couple of years. Canada doesn't yet have any offshore wind farms, because they are more expensive and there's still enormous room for the growth of land-based wind farms.

In 2003, Canada had 12 megawatts of installed solar PV capacity. By 2013, that number had leapt to 1,200 megawatts, one hundred times higher. Ninety percent of that leap was made in the most recent five-year period. Ontario's Green Energy Act, modelled after German and Japanese legislation, has been a major impetus for the growth of a Canadian solar industry. Large solar PV plants are under construction in British Columbia and Ontario, while Canada's first concentrating solar power plant opened in Medicine Hat, Alberta, in 2014.

Other sources of renewable energy are lagging behind in Canada. Nova Scotia has a tidal power plant with a generating capacity of 20 megawatts and is trying to harness the incredible power of the Bay of Fundy's tides. The first turbine deployed in the Bay of Fundy was badly damaged and had to be removed just months after it was installed in 2009. James Ives, CEO of OpenHydro, admitted, "We underestimated the energy in the Bay of Fundy." In B.C., there is a small experimental wave energy project. Despite tremendous potential, Canada has zero megawatts of geothermal electricity generation, trailing international leaders including the U.S., the Philippines, Indonesia, Mexico, Italy, New Zealand, Japan, Kenya, and Costa Rica. Accelerating Canada's transition to renewable energy in order to meet our greenhouse gas reduction targets under the 2009 Copenhagen Accord would create 620,000 person years of construction jobs and 34,000 permanent jobs by 2022.

Professors Mark Jacobson and Mark Delucchi have published a detailed analysis illustrating how the U.S. could shift to 100 percent renewable energy for all its needs—residential, commercial, industrial, and transportation—by 2050. Their conclusions have been confirmed by American government agencies and independent research institutes. The U.S. National Oceanic and Atmospheric Administration indicated that 80 percent of American electricity needs could be generated by renewables by 2030. The resulting system would produce 90 percent less greenhouse gas emissions and cut water use in half. Wind and solar would dominate, with smaller contributions from geothermal, hydropower, and biomass. The study did not include enhanced geothermal systems, wave and tidal power, or offshore wind turbines. American renewable energy potential is estimated to be roughly 128 times the generating capacity of the current U.S. electrical system.

The shift to renewables is underway. Wind power in the U.S. tripled between 2007 and 2012. By 2014, the amount of installed solar capacity was 15 times higher than in 2008. In the first quarter of 2014, more than 90 percent of new electricity generating capacity in the U.S. came from renewable sources, primarily solar and wind. Over 175,000 people are employed in the American solar industry, and there are more than 625,000 renewable energy jobs in the U.S. As President Obama noted in 2014, "Every four minutes, another American home or business goes solar." In 2014, greenhouse gas emissions from the U.S. power sector fell to their lowest level in 20 years.

The U.S. is in the middle of a project called the SunShot Initiative, intended to be the 21st-century equivalent of putting a person on the moon. The goal is to make solar energy fully cost-competitive with traditional energy sources by 2020 by lowering the cost of solar powered electricity to six cents per kilowatt-hour (kWh), which is less than anyone in North America currently pays for electricity provided by a utility company. Less than four years after it started, the project was already two-thirds of the way towards meeting its goal, with unsubsidized solar costs falling from 21 cents/kWh to 11 cents/kWh. The cost of wind energy is also plummeting, having dropped 43 percent in the U.S.

since 2009—making it now cheaper than coal for new generating capacity in some locations.

The U.S. is the world leader in geothermal energy. Geothermal electricity is an important element in the renewable energy mix because it provides consistent baseload power, unlike intermittent sources such as wind and solar. Geothermal requires a small area of land, small volumes of water, and produces a tiny fraction of the carbon dioxide emissions of fossil fuel power plants. Capital costs are relatively high, but the fuel cost is zero. El Salvador, Kenya, the Philippines, Iceland, and Costa Rica already generate more than 15 percent of their electricity from geothermal sources. A large plant (140 megawatts) in Kenya that began operating in 2014 boosted geothermal's share of electricity production to over 50 percent and led to a 30 percent reduction in electricity costs. By providing electricity to thousands of schools, it also facilitated a free laptop program, creating unprecedented opportunities for Kenyan students.

Because of the size of its economy, China is one of the key players in the future of the planet's climate. China leads the world in some unenviable categories, including total electricity consumption, coal production, coal imports, coal consumption, carbon dioxide emissions, and worst air quality. Public protests against proposed coal plants and waste incinerators are increasingly common, and some protests have turned violent. In a town called Zhongtai, police vehicles were torched and 39 people were injured in an antipollution demonstration in 2014. Premier Li Keqiang vowed to decrease pollution and described the need to create "an ecological civilization."

Positive changes are underway. China already leads the world in hydroelectricity production, although massive dams have significant social and environmental costs. China also leads the world in solar power, wind generation, and clean energy investment. In 2013, for the first time, Chinese investment in clean energy exceeded money spent on new fossil fuel power plants, and new renewable power capacity exceeded new coal capacity. In 2014, for the first time, Chinese demand for coal declined relative to the previous year. More than 2.6 million Chinese are employed in the renewable energy sector. Chinese leaders are beginning to place restrictions on the burning of coal, closing down some smaller facilities, and prohibiting the construction of new coal plants in or near large cities (where air pollution problems are the most grave). These actions have the potential to save millions of lives and are vital elements of the global effort to reduce the greenhouse gas emissions that are causing climate change.

Australia is a sunny continent that has historically relied on coal to generate the lion's share of its electricity. But in just five years, from 2008 to 2013, the number of rooftop solar PV systems in Australia soared from 8,000 to 1.1 million, fuelled by government subsidies. In the states of Queensland and South Australia, one in four homes now has a solar PV system on its roof. The Australian PV industry employs 17,000 people in full-time jobs. A government study published in 2013 indicated that the transition to 100 percent renewable electricity by 2030 would be on par economically with the investments needed to maintain a system based predominantly on fossil fuels. Other studies show solar and wind power are now cheaper than coal and gas in Australia, even before health and environmental externalities are factored into the comparison. For example, wind energy was pegged as 14 percent cheaper than new coal and 18 percent cheaper than new natural gas. "The perception that fossil fuels are cheap and renewables are expensive is now out of date," said Michael Liebreich, chief executive of Bloomberg New Energy Finance. Liebreich added, "The fact that wind power is now cheaper than coal and gas in a country with some of the world's best fossil fuel resources shows that clean energy is a game changer which promises to turn the economics of power systems on its head."

A list of countries where more than 90 percent of the electricity supply comes from renewable energy will raise a lot of eyebrows. Iceland combines hydro and geothermal to get to 100 percent. Despite the substantial environmental impacts of hydroelectricity, it generally creates far fewer greenhouse gas emissions than fossil fuels. As noted earlier, all types of energy have impacts, which is why the first priority in the world's wealthy nations needs to be using less, not generating more. Albania and Paraguay generate all of their electricity from hydro. Costa Rica (hydro, geothermal, and wind) and Norway (hydro, wind, biomass, and solar) are very close to achieving their target of 100 percent. Small islands from Samsø (Denmark) to El Hierro (in Spain's Canary Islands off the coast of Africa) have successfully switched from burning diesel for electricity to a mix of 100 percent renewable sources. By relying primarily on hydro, ten other countries have surpassed 90 percent: Belize, Bhutan, Burundi, Democratic Republic of Congo, Kyrgyzstan, Laos, Mozambique, Nepal, Tajikistan, and Zambia. A growing number of companies, including Apple, IKEA, and even Walmart, have pledged to be 100 percent powered by renewable energy.

Other countries have set audacious targets for shifting to clean energy. Scotland and Uruguay are striving for 100 percent by 2020, New Zealand 90 percent by 2020. Germany has a legally binding target to produce at least 35 percent of its electricity from renewable sources by 2020—with the target rising to 50 percent by 2030, and 80 percent by 2050. Since 2000, Germany has quintupled the portion of electricity generated by renewable sources, going from 6 percent to 30 percent. Arne Jungjohann, program director for the environment at the Heinrich Böll Foundation, a German think tank, said, "Today, you see windmills across the country, blue shining solar arrays on rooftops and town halls." Half of all wind projects are community owned by small-scale investors and farmers who, according to Jungjohann, "invest their money in a wind park instead of in the bank." On one day in May 2014, wind and solar produced a record 75 percent of the total electricity supply. This progress was propelled primarily by innovative government policies, including a feed-in tariff mandating generous subsidies for wind and solar. First enacted in 1991 and adjusted several times since, Germany's feed-in tariff gives renewable energy projects priority access to the power grid and provides a guaranteed price for the power for up to 20 years. Although Germany's experience has generated controversy about its economic costs, the OECD calculates that there has been a tiny negative impact on Germany's GDP (approximately 0.2 percent). The increased cost to consumers' electricity bills averages about $280 per year. Jobs in Germany's renewable energy industry have more than tripled over the past decade to 371,000 in 2013 and are projected to grow to 500,000 by 2020.

Other countries are also making tremendous progress. Wind supplied 30 percent of Denmark's electricity in 2012. The Danish goals are to produce 50 percent of its electricity from wind by 2025 and 100 percent of its electricity from renewable energy by

2050. Wind power is now the leading source of electricity in Spain, where reliance on coal, natural gas, and nuclear has simultaneously dropped. In 2013 alone, greenhouse gas emissions from the Spanish electricity sector fell by 23 percent. Portugal generates two-thirds of its electricity from renewables, with wind and solar growing rapidly. In the aftermath of the Fukushima nuclear disaster, the Japanese government moved quickly to expand its feed-in tariff program for renewables. Over 4 gigawatts of new solar electricity generating capacity were added in the first year, the equivalent of four nuclear reactors. Another 25 gigawatts of clean energy projects (mostly wind and solar) have been approved.

Just as people in developing countries went straight from no phones to mobile phones, leap-frogging over costly landline infrastructure, so too are some nations bypassing conventional electricity systems. The shift to solar could be an incredible boost to hundreds of millions of people who currently live with no electricity in their homes. In south Asia, Africa, and Latin America, communities are adopting locally generated solar electricity instead of waiting for power from a huge, distant power plant fuelled by coal, natural gas, or uranium. The savings associated with lower infrastructure costs are immense. In 2014, India's new prime minister, Narendra Modi, launched an ambitious initiative to bring solar electricity to 400 million people by 2020, scaling up a successful program he oversaw in his home state of Gujarat. A similar program already exists in Peru, where the government will spend more than $200 million on small-scale solar PV systems by the end of 2016 to ensure 95 percent of households have access to electricity. Peru's energy minister Jorge Merino said, "This program is aimed at the poorest people, those who lack access to electric lighting and still use oil lamps, spending their own resources to pay for fuels that harm their health."

In poor but sunny Bangladesh, two new rooftop solar PV systems are installed on homes and businesses every minute, for a total of more than one million systems annually. This rapid progress is ascribed to falling prices and the widespread availability of micro-credit programs. According to Munawar Misbah Moin, managing director of one of Bangladesh's leading solar companies, a typical small PV system costs roughly $300, with customers making down payments of $45 and monthly payments of $8–10 for two to three years. These small systems can power lights, a cell phone charger, and a small television. The impact on quality of life, Moin says, is "unbelievable" when families gain access to electricity for the first time.

In many African countries, solar power is cheaper than burning imported diesel fuel, which is often the primary source of electricity. Canadian companies Windiga Energy and JCM Capital are developing clean energy in Africa, with major solar projects planned, underway, or completed in Burkina Faso, Cameroon, Ghana, Malawi, and Nigeria. A single 20 megawatt solar power plant in Burkina Faso will add 10 percent to the installed capacity of the country's grid. According to Windiga CEO Benoît Lasalle, diesel power costs 35 to 38 cents per kilowatt hour in West Africa, whereas solar costs 21 cents per kilowatt hour, leading him to say, "You do the math."

THE PROGRESS DESCRIBED in this chapter is extraordinary. Solar is growing exponentially, with increases of more than 40 percent per year over the past decade and a half. The share of wind, solar, geothermal, and bioenergy in global electricity production doubled from 2 percent to 4 percent between 2006 and 2011, and will likely continue to double every five or six years. In 2012, 2013, and 2014, renewable energy accounted for more than half of the new electricity-generating capacity added globally. The International Energy Agency now projects that renewable energy will supply more electricity to the global grid than natural gas by 2016, and if the right policies are put in place, solar could be the world's largest source of electricity by 2050. In light of the IEA's track record of underestimating renewables, these are remarkable predictions.

Challenges related to grid access, intermittency, and storage are being addressed. Today's renewable energy technologies, which are certain to continue improving and becoming cheaper, are already capable of providing 100 percent of the world's electricity needs within the relatively short span of a few decades. Even Shell, the notorious oil giant, expects solar to be the world's number-one source of energy by 2100. Although some critics attack subsidies granted to renewable energy, the reality is that these pale in comparison to the subsidies still given to the fossil fuel industry. In 2013, oil, gas, and coal companies received $550 billion in subsidies compared to $120 billion for all renewables combined. In the editorial of a 2015 issue dedicated to the transformation of the world's energy supply, the *Economist* described fossil fuel subsidies as "nonsense" and throwing money down a "rathole," and urged governments to eliminate them all.

In 2013, investors put $254 billion into renewable energy, led by China, the U.S., Japan, the U.K., and Germany. This level of investment needs to be rapidly ramped up towards a trillion dollars annually in order to avoid the apocalyptic climate change scenarios that scientists are predicting under a business-as-usual scenario, such as a 4°C rise in global average temperature by the end of this century. Another factor favouring investment in renewable energy is a 2014 study of hundreds of electricity projects in 57 countries between 1936 and 2014 that found cost overruns and construction delays are much less likely and of far smaller magnitude for solar and wind projects compared to nuclear, hydroelectric, and fossil fuel projects.

The International Energy Agency estimated that it would cost $44 trillion to switch the global energy system to 100 percent renewables by 2050 but would save $115 trillion in fuel costs, for a net savings of $71 trillion. This shift would also prevent more than three million premature deaths and tens of millions of illnesses annually, and diminish conflicts over increasingly scarce fossil fuel resources. When the health and environmental externalities of burning fossil fuels are taken into consideration, the economic rationale for rapidly switching to a renewable energy system is irrefutable. As Adnan Z. Amin, head of the International Renewable Energy Agency, said at the release of a major report on plummeting renewable costs in 2015, "It has never been cheaper to avoid dangerous climate change, create jobs, reduce fuel import bills, and future-proof our energy system with renewables."

"We would accomplish many more things if we did not think of them as impossible."
—Vince Lombardi

Chapter 3
The Circular Economy

A PAIR OF RUBBER BOOTS may seem like an unlikely beginning for a chapter about the transformation of the modern industrial economy. On Canada's wet west coast, a good pair of wellingtons is a necessity of life, and constitutes acceptable footwear at every sort of function, from dinner parties to classical music concerts. One day in Pender Island's hardware store, I discovered a pair of ordinary looking boots with an extraordinary feature. When the boots wear out, which is inevitable in this blackberry- and barnacle-laced terrain, you simply return them to the manufacturer to be melted down and made into new rubber boots. This circular process mimics the billion-year wisdom of natural systems in which waste equals food.

Today we are living within a linear economic system that has outlived its usefulness. We extract resources, make products, and then throw them away, generating waste and pollution at each stage. Humans are consuming resources faster than ecosystems can replenish them, and our waste and pollution exceed Nature's ability to absorb them. Our collective ecological footprint would require 1.5 planets like Earth to sustain us. The consequences of our excess include climate change, extinctions, and the accumulation of persistent organic pollutants in wildlife and ecosystems from North Pole to South Pole. To make matters worse, the global population is spiralling inexorably towards 9 billion in 2050. At the same time, billions of people are leaving poverty behind and entering the middle class. With the resultant surge in demand for consumer goods, it should be blindingly obvious that the linear economy needs to be replaced as soon as possible. Regardless of whether particular resources are actually running out, the outlook is one of supply disruptions, volatile commodity prices, accelerated environmental degradation, and rising political tensions.

Fortunately, Nature offers an impressive and inspiring blueprint for redesigning today's economy. In living systems, one species' waste is another species' food. Materials are recycled indefinitely, which is an essential part of the definition of sustainable. An amazing example from B.C. involves Pacific salmon, which spend years growing in the ocean before returning to coastal rivers and streams to spawn. The salmon are caught by bears, wolves, ravens, and other predators, who carry the carcasses into the forest. The remains of the salmon nourish the soil and contribute to the growth of riparian vegetation and towering old-growth forests, which in turn provide benefits for aquatic habitat. The planetary cycles of water, oxygen, and nitrogen also demonstrate the way natural systems are designed to carry on in perpetuity.

Visionaries have seen that emulating Nature as a model for a new economic system is one of the most promising avenues for liberating us from the resource constraints and environmental problems that threaten to undermine the progress of the past centuries. The circular economy (also described as a one-planet or biosphere economy) recognizes that the Earth is finite—that there are physical limits to non-renewable substances like fossil fuels and minerals, and limits to the volume of renewable resources such as fish and forests that ecosystems can produce. There are also limits to Earth's capacity to assimilate our waste, which we exceed at our peril.

In contrast to today's linear economy, a circular economy recycles materials infinitely, relies on renewable energy, minimizes and eliminates the use of toxic chemicals, and eradicates pollution and waste through careful design. Inspired by living systems, it provokes new designs, materials, systems, processes, and business models. Redesigning products and systems will transform today's wastes into tomorrow's useful and affordable raw materials, reducing pressure on the environment. For example, a growing number of cities are collecting residents' food scraps and yard waste and processing these "wastes" into compost, heat, and liquid fuels for transportation. Instead of going to the dump, these "wastes" are turned into valuable resources. The idea of a circular economy is not new. In 1966, American economist Kenneth Boulding wrote an extraordinary essay in which he called for a shift away from "the cowboy economy," where endless frontiers imply no limits on resource consumption, pollution, or waste, to a "spaceship economy," where everything is engineered to be constantly recycled. There are several principles at the heart of the circular economy, including cradle-to-cradle design, zero waste, and biomimicry (innovation inspired by Nature).

MICHAEL BRAUNGART AND William McDonough are the visionaries behind cradle-to-cradle design, a radical new approach that strives to eliminate the concept of waste. Braungart is a German chemist and former Greenpeace activist. McDonough is an American architect. They met, by chance, in Manhattan in 1995 and their first conversation sparked a collaboration that has produced new ideas, best-selling books, and hundreds of innovations in fields ranging from textile manufacturing to green building. As they state in their 2013 book, *The Upcycle: Beyond Sustainability—Designing for Abundance,* "Human beings don't have a pollution problem; they have a design problem. If humans were to devise products, tools, furniture, homes, factories, and cities more intelligently from the start, they wouldn't even need to think in terms of waste, contamination, or scarcity . . . Good design would allow for abundance, endless reuse, and pleasure."

One of their first projects was to design a fabric for Steelcase, a furniture company that wanted a greener product. Braungart and McDonough reviewed over 8,000 chemicals commonly used in the textile industry and identified only 38 that did not harm either human health or the environment. Using those 38 for dyes and finishes, they designed an award-winning line of fabrics. When inspectors tested the effluent leaving the Swiss mill where the fabric was produced, they thought their instruments had malfunctioned. The water leaving the mill was as clean as the water going in. Trimmings from previous fabrics had been classified as hazardous waste and could not be burned or buried in Switzerland due to their chemical content, but trimmings from the new product are donated to gardeners to use as mulch. The cradle-to-cradle idea took off, and Braungart and McDonough created a certification system to help guide the redesign process and ensure that products using the name lived up to their claims.

Cradle-to-cradle design rethinks industrial and consumer products so that they can be made from materials that are either 100 percent recyclable or 100 percent compostable. When a product reaches the end of its useful life, it has to be capable of being remanufactured into new products (the technical cycle), composted to produce soil to grow new crops (the biological cycle), or easily separated into these two categories. The technical cycle consists of products that can be repeatedly recycled, such as glass, metals, or my rubber boots. The biological cycle consists of materials that can be safely composted and added to soil to grow food, trees, or other useful crops. Toxic materials, emissions, and effluents are eliminated through careful design and engineering. Among the myriad products that already have succeeded in the rigorous cradle-to-cradle certification process are carpets, furniture, clothing, shampoo, and building materials.

For example, the comfortable chair that I'm sitting in as I write this book is an Aeron, a cradle-to-cradle product from Herman Miller, an American office furniture company. The Aeron chair earned silver status from McDonough and Braungart's certification process. Its aluminum components are made from 100 percent recycled material and can be recycled again. Steel components contain approximately 25 percent recycled content and are 100 percent recyclable. Aeron chairs were redesigned to eliminate the use of polyvinyl chloride (PVC), a toxic substance known to cause cancer in factory workers. Metal components are coated with a paint that emits negligible volatile organic compounds (VOCs). Plastic components are identified with a code, to make them easier to recycle. The seat frame and back contain over 60 percent recycled content, made from approximately 36 recycled two-litre plastic beverage bottles per chair. Foam and textile materials can be recycled into everything from automotive components to carpet padding. Packaging materials are 100 percent recyclable, including corrugated cardboard and a polyethylene plastic bag. The Aeron chair is also certified as a low-emitting product that meets indoor air quality standards. I expect that it will last my entire life, and likely serve as Meredith's chair for decades after I'm gone. None of this is rocket science. It's just good design—manufacturing a high-quality chair while also thoughtfully considering the eventual end of its useful life.

Carpet manufacturers are another unlikely source of inspiring examples about the potential for redesigning ordinary products to eliminate toxic substances and achieve extraordinary reductions in waste and pollution. Interface Carpets, Shaw Industries, and Desso are companies that have redesigned their products to be either completely recyclable, biodegradable, or easily separated into these two streams. Interface's ambitious environmental strategy, Mission Zero, aims to eliminate the company's negative environmental impacts by 2020. Interface invented a glue-less installation process (designing out the use of toxic volatile organic compounds), eliminated the use of heavy metals, and recycled over 100 million pounds of carpet that would otherwise have ended up in landfill sites. Over the course of a decade, Interface reduced waste by 94 percent, energy use 39 percent, greenhouse gas emissions 71 percent, and water use 83 percent. In addition to these environmental benefits, Interface saved hundreds of millions of dollars, won industry design awards, and boosted sales to record levels.

Inspired by the ideas of a circular economy and zero waste, Kingfisher, Europe's largest home improvement retailer (their stores include B&Q and Screwfix), launched a program called Net Positive. In the first year of the program, 2013–2014, the Screwfix chain successfully achieved the goal of sending nothing to landfill, by diligently working with a recycling company to come up with waste solutions. For example, Screwfix manufactured a kitchen counter made from paper waste and created packaging for bedding plants that is 100 percent recyclable. By 2020, Kingfisher aims to offer 1,000 cradle-to-cradle products.

A comprehensive study on cradle-to-cradle certification recently found both environmental and economic benefits. Companies in the report include AGC Glass Europe, Aveda, Construction Specialties, Ecover, Mosa, Puma, Shaw Industries, Steelcase, and Van Houtum. These businesses enjoyed reduced costs, improved product value, increased profit margins, new revenue streams, and avoided risks such as volatile commodity prices and supply disruption. The environmental benefits included less use of toxic materials and more efficient use of energy, water, and other resources. Employees benefitted from healthier workplaces and opportunities to contribute to sustainable enterprises. Since McDonough and Braugart's program began in 2005, over 200 companies worldwide have participated in the cradle-to-cradle certification program, with hundreds of product lines representing thousands of different products certified.

THE SANDWICH ME IN RESTAURANT in Chicago demonstrates how the notoriously wasteful food industry can virtually eliminate waste through intelligent design and planning. Owner-operator Justin Vrany strives to serve good food, made in-house, while setting a zero waste goal. Local farms provide minimally packaged food. Almost everything gets reused or repurposed. Vrany explained that he practices the five R's, adding reject and refuse to the usual trio of reduce, reuse, and recycle. At Sandwich Me In, food scraps are returned to farms for composting or chicken feed. A single plastic bag of garbage accumulated over the course of two years, such as plastic-lined paper coffee cups left behind by customers. The Parker restaurant in Vancouver, serving gourmet vegetarian fare, similarly produces less than one pound of garbage each month. And the Loblaws grocery store chain is now reducing food waste by selling "ugly" fruits and vegetables at a discount, following the successful example set by Intermarche in France.

Another striking example of taking what was previously regarded as waste and repurposing it into a valuable resource comes from the coffee industry. When coffee is grown, the beans are contained in a husk known as the coffee cherry, which usually is discarded, creating vast volumes of waste. It was recently discovered that the cherries can be ground into fine "coffee flour," a product that has myriad uses in baking and cooking, is gluten-free, and is packed with nutrients. Coffee flour has five times more fibre than whole-wheat flour, three times the protein of kale, three times as much iron as spinach, and an ounce of coffee flour has twice the potassium of a whole banana. Seattle chef Jason Wilson raves about the flavour this product adds to dishes from fresh pasta to coffee cake, when used as a partial substitute for all-purpose flour. According to Wilson, "From flavour to consistency to backstory, coffee flour is the rarest thing I've ever cooked with." Coffee flour has the potential to solve an environmental problem, help feed a hungry world, and bolster the income of coffee farmers.

ANOTHER CORNERSTONE OF THE circular economy involves the nascent field of biomimicry, literally imitating life. Janine Benyus, author of *Biomimicry: Innovation Inspired by Nature*, says this new field "studies Nature's best ideas and then imitates these designs and processes to solve problems." Over the course of 3.8 billion years, Nature has worked out solutions to many of the problems facing humanity today. Biomimicry already has an impressive catalogue of success stories. One of the earliest examples was the development of Velcro. In the 1940s, Swiss engineer George de Mestral noticed that his socks and his dog's coat were often covered in burdock burrs after hiking in the Alps. De Mestral put a burr under a microscope and discovered a simple design of hooks superbly suited for attachment. He experimented for years before figuring out how to achieve the same effect, inventing Velcro and patenting it in 1955.

When Japan's high-speed bullet trains were first deployed, pressure waves built as the trains travelled through narrow tunnels, creating loud sonic booms at the exit. Facing noise complaints, the manufacturer challenged its engineers to improve the design. Eiji Nakatsu was one of those engineers, and also a passionate birder. He redesigned the front of the train to resemble a 50-foot-long kingfisher's beak. These delightfully gregarious birds have streamlined beaks, specially adapted so they can enter the water when fishing without making a splash. Nakatsu also modified the train by adding many small vortices to the pantograph, a protrusion that extends above the train to receive electricity from overhead wires. These changes mimicked a feature of the feathers of owls that enables them to fly silently and thus hunt successfully. These Nature-inspired improvements not only made the bullet trains quieter, but also increased their speed by 10 percent and reduced electricity use 15 percent.

More recently, Nature has served as the basis for a life-saving medical innovation. One of the major challenges facing health authorities in developing nations is that vaccines need to be kept refrigerated, a difficult task in areas unserved or unreliably served by electricity. As many as half of all vaccines can be wasted because of breaks in the chain of refrigeration between production and patient. To the rescue came biomimicry. An ancient species called tardigrades, a tiny cousin of today's arthropods, is able to dry out for as many as 120 years without dying. A process called anhydrobiosis protects the creature's critical chemical systems—DNA, RNA, and proteins—until they are exposed to water and revived. Inspired by tardigrades, a company called Biomatrica has created a process that shrink-wraps vaccines so that refrigeration is unnecessary.

Geckos are small lizards that can scale smooth walls and sprint upside-down across ceilings. Their powerful grip is created by millions of microscopic hairs on the bottom of their toes. While each hair's connection to any given surface is minuscule, the cumulative effect is impressive. Inspired by these little lizards, a team of University of Massachusetts researchers developed Geckskin, an adhesive so strong that a strip the size of an index card can hold up to 700 pounds on a smooth surface. A form of gecko tape could replace sutures and staples in the hospital. The ability to don a pair of gecko-inspired gloves and socks so that you can scale walls like Spiderman may not be far off.

Other examples of biomimicry in action include water collection systems inspired by the Namibian beetle, Mercedes-Benz vehicles shaped like a tropical fish, and improvements to LED lights based on the anatomy of fireflies. Volvo is developing collision avoidance technology based on locusts. Hospitals are adopting a product based on the skin of sharks that prevents bacteria from adhering to surfaces. Calera, a company inspired by coral reefs, is using carbon dioxide to make cement. Whale Power developed wind turbines with scalloped edges that reduce drag, based on the fins of humpback whales. Scientists are trying to emulate trees, which efficiently pump water upwards from roots to branches and leaves, and Amazon eels, which can produce up to 600 volts of electricity. The possibilities for effective redesign of products that reduce energy use and waste based on wonders of the natural world are almost endless.

EXPERTS HAVE IDENTIFIED five business models based on circular economy principles: circular supplies, resource recovery, products as services, product life extension, and sharing platforms. Circular supply companies provide renewable energy, bio-based (compostable) materials, or fully recyclable substances such as aluminum to replace fossil fuels and single-use materials. Resource recovery firms find value in recovering or repurposing materials previously considered wastes or by-products. Retailer H&M collects used clothing in its stores to remake into new garments, closing the loop in textile production. Companies that offer services instead of products shift incentives so that longevity and recyclability are top priorities. Carpet companies like Interface and Desso are using this approach, leasing carpets instead of selling them, while Philips is doing the same with lighting and Michelin with tires (fleet customers pay for miles driven instead of owning the tires). Product life extension focuses on making items durable instead of disposable. Some motor vehicle companies remanufacture parts that meet the same performance standards and carry the same warranty as new parts but without new materials and at half the cost. European carmaker Renault has achieved reductions of 80 percent for energy, 88 percent for water, and 77 percent for waste by remanufacturing rather than making new components. Gazelle, the leading web-based electronics recycling company in the U.S., buys and resells used gadgets, keeping millions of phones and tablets out of the waste stream.

Sharing platforms encourage the increased use of products that otherwise are underutilized. Collaborative business models are on the rise, such as car-sharing through companies like Autoshare or Zipcar, smartphone based ride-sharing such as Uber and Lyft, or renting out private homes or rooms through Airbnb and its competitors. ParkFlyRent is a business that allows people to park at airports and earn income while they are away by renting their car to others. These new enterprises are the beginning of an emerging sharing economy that, according to the founders of Gazelle, "changes the way we all think about buying, owning, selling, and recycling." Sharing also works on a non-business model too, as libraries have long demonstrated. The library model is being expanded to include items ranging from tools to kitchen equipment. Our Pender Island community has an apple juice press and a food dehydrator that all residents are welcome to borrow. Like many small communities, we have a thrift store (called the Nu-to-Yu) that collects and resells items ranging from toys to clothing. Although Pender's population is only 2,500 full-time residents, the Nu-to-Yu has donated over $1 million to island charities and projects. Websites that facilitate the sale of second-hand merchandise, from eBay to Criagslist, are making it easier than ever to buy and sell used goods. A study released in 2015 reported that Canadians spend over $30 billion on secondhand goods annually.

The leaders in circular economy innovation are Europe (at both the EU and national levels), Japan, and China. Europe recognizes that current levels of consumption must be reduced to offset the burgeoning environmental pressures created by the economic growth of developing countries. In 2012, the European Commission stated, "In a world with growing pressures on resources and the environment, the EU has no choice but to go for the transition to a resource-efficient and ultimately regenerative circular economy." In proposing a set of circular economy targets for 2030, European Environment Commissioner Janez Potočnik said,

> We are living with linear economic systems inherited from the 19th century in the 21st-century world of emerging economies, millions of new middle class consumers, and inter-connected markets. If we want to compete, we have to get the most out of our resources, and that means recycling them back into productive use, not burying them in landfills as waste. Moving to a circular economy is not only possible, it is profitable, but that does not mean it will happen without the right policies.

Under European Commission proposals, single-use, non-recyclable products and many toxic substances would be banned. Sending compostable or recyclable material to landfill also would be banned. Europe is striving to recycle 70 percent of municipal waste and 80 percent of packaging waste by 2030. And counter to many fears, meeting these goals would generate roughly 580,000 new jobs. A study funded by the Ellen MacArthur Foundation, a think tank dedicated to this topic, concluded that applying the circular economy framework to a subset of European manufacturing sectors could result in $630 million in annual savings for firms in those industries.

Recognizing the Earth's limits requires that we fundamentally change our resource consumption. Scandinavian nations are the world's leaders in this regard. Sweden acknowledges, "It is becoming ever more apparent that, when the consumption patterns of the industrialized world are adopted by the developing countries, the aggregate extraction of resources worldwide will exceed planetary limits." Norway recognizes that "if environmental limits are exceeded, the consequences may be irreversible . . . Norway's sustainable development policy, including its environmental protection policy, is therefore based on environmental targets that take account of the tolerance limits of the environment."

Denmark, Scotland, Germany, and the Netherlands are leading European nations in promoting the shift from a linear to a circular economy. Denmark is home to one of the world's best examples of industrial symbiosis, where the "waste" products from one corporation serve as raw materials for another. At the Kalundborg industrial complex, eight companies and the municipality trade steam, ash, gas, heat, sludge, water, and other materials. Resources are reused and conserved instead of being wasted, the environment benefits, and the cost savings are substantial. Scotland has made progress on its transition to a circular economy through its ambitious plans to achieve zero waste and, by 2020, a 100 percent renewable electricity grid. In 2014, Scotland announced the creation of a Scottish Institute for Remanufacture and a Scottish Materials Brokerage Service to expedite the shift to a circular economy.

As a country with limited natural resources, Japan has a long record of striving for efficient manufacturing. In 2000, Japan passed the pioneering Law for the Promotion of Effective Utilization of Resources, requiring manufacturers to treat goods as circular products, run disassembly plants, and recover materials. As a result, 98 percent of metals are recovered from waste streams so that they can be reused.

In 2009, Chinese leaders, increasingly concerned about the environmental degradation inflicted by their country's meteoric economic growth, passed a law called the Circular Economy Promotion Law. The law's objectives include decoupling economic growth from resource consumption and pollution, shifting from a narrow vision of waste management to the idea of closed-loop material flows, and selling services instead of products. In the latest version of their circular economy plan, China pledged to limit economic growth and place limits on their consumption of water, land, energy, and resources, as well as pollution.

The China Association for a Circular Economy—a group of governments, researchers, and businesses—claims that China's circular economy is growing by more than 15 percent annually, and already is measured in the hundreds of billions of dollars. Peter Lacy, head of sustainability for the management consulting firm Accenture in the Asia-Pacific, estimated that between six and seven million jobs will be created in the next decade by the circular economy in China as a result of what he calls disruptive innovation. Goodbaby, a Chinese company that is the world's largest manufacturer of strollers, has been developing cradle-to-cradle products since 2010. The Shandong Tranlin Group, a paper manufacturer, has developed a technology that can process straw for use in its paper products and upcycle wastes from this process into organic fertilizer. Eco-cities, eco-provinces, and eco-industrial parks are being planned using the principles of the circular economy.

THE CIRCULAR ECONOMY INITIATIVES described in this chapter have major strategic importance for the whole world. They mean the end of our take, make, throw away method of business, which has become an unsustainable economic model. In natural systems, the quantity and quality of "wastes" are such that they can all be used by other organisms, resulting in an inherently sustainable process powered by the sun's energy. By intelligently redesigning systems of production and consumption, using Nature as a model, humans can ensure that our "wastes" are also of a quality and quantity capable of being recycled into something useful. Consumers need to think in new ways about the things they need and be open to leasing or renting instead of owning, enabling manufacturers to build new business models and be responsible for the durability, reuse, and recycling of their products. Governments need to enact stronger laws governing the disposal of waste, the manufacturing of disposable products, and the elimination of toxic substances. Businesses need to rethink their current approach and embrace the opportunities offered by the circular economy, cradle-to-cradle design, and biomimicry. Hundreds of companies are already doing so and flourishing as a result. Disruptive technologies—solar, smartphones, the Internet, software, nanotechnology, computer processing speeds—are facilitating these shifts. Investment, both public and private, in game-changing technologies is also essential. According to the Ellen MacArthur Foundation, adoption of the circular economy could yield trillions of dollars in resource savings annually, along with enormous benefits for human and ecosystem health.

Part II
HEALTHY ENVIRONMENT,
HEALTHY PEOPLE

"The future belongs to those who believe in the beauty of their dreams."
—Eleanor Roosevelt

Chapter 4
Cleaner Air

THANKS TO GOOGLE, it has never been easier to time travel. You can sit at home and pull up historical photographs, newspapers, and magazine articles from a particular time and place. Focusing on air pollution demonstrates dramatically how far and how quickly progress has been made in North America, Europe, and some leading Latin American cities. At the same time, today's nightmarish photographs and video from cities in China, India, and Africa show the urgent need for others to emulate our success. To illustrate how far we've come in cleaning up the air we breathe, let's visit Los Angeles and London, some 60 years ago.

Newspaper photographs from the 1950s in Los Angeles show big cars, big hairdos, and the frequent impossibility of seeing across the street. Air pollution from the ballooning number of vehicles on the road and emissions from coal-fired electricity plants turned the City of Angels into a smoggy version of hell. The sky was so dark on summer days in downtown L.A. that some people thought there was a solar eclipse. On those bad days, everyone who ventured outside experienced itchy eyes and sore throats or, more seriously, asthma attacks and chest pain. The auto industry vehemently denied any connection between cars and smog. In 1953, Ford argued that automobile exhausts "are dissipated in the atmosphere quickly and do not present an air pollution problem." By the late 1950s, Ford claimed that the smog problem was "not bad enough to warrant the enormous cost and administrative problems of installing three million afterburners" (which filter exhaust before it exits the tailpipe). But by 1968, the evidence had mounted, and the U.S. Department of Justice laid charges against Ford, Chrysler, General Motors, and the Automobile Manufacturers Association for a lengthy conspiracy to prevent and delay the introduction of pollution control devices for motor vehicles. Jack Doyle, author of *Taken for a Ride: Detroit's Big Three and the Politics of Pollution*, observed, "Time and time again in the 1950s, 1960s, 1970s, 1980s, and through the 1990s, the automakers said, 'we don't have the technology,' 'it's impossible,' 'we don't have the money,' 'we don't have the engineers,' 'we're at a competitive disadvantage,' 'jobs will be lost,' 'it will take ten years,' 'it will make cars unsafe,' etc." As late as 1981, then U.S. President Ronald Reagan stunningly claimed, "Trees cause more pollution than automobiles." (To the contrary, trees and other terrestrial plants generate half of the oxygen we breathe, while absorbing carbon dioxide.) In their book *Pollution and Policy*, historians James Krier and Edmund Ursin conclude, "The industry was remarkably slow in conceding what had become obvious to everyone else—the importance of automotive emissions."

London was the largest city in the world in 1952, home to eight million of the world's 2.6 billion people. And on December 5,

London was struck by an air pollution disaster.

A cold front moved westward from the European mainland and settled over the city like a blanket. At the time, most people still heated their homes with open coal fires. Factories also burned coal. The coal smoke rose as usual that day, but after hitting the cold air it cooled and fell back to earth, creating a suffocating sulphurous haze. In some parts of London, visibility dropped to less than a metre. Traffic came to a standstill, and vehicles belching exhaust exacerbated an already ugly situation. Conductors on foot, carrying flares, led buses through the smog, while police with burning torches directed traffic. The cast of *La Traviata* called off their performance of the opera because the audience could no longer see the stage—inside the theatre! Football games were cancelled because it was impossible to see the ball.

The impenetrable smog lasted five days, causing at least 4,000 premature deaths. Undertakers and funeral homes ran out of coffins. Florists ran out of flowers. In the next three months, an additional 12,000 people died, hospital admissions for respiratory diseases jumped 160 percent, and cases of pneumonia tripled. Headlines in the *Guardian* read "Busy time for thieves" and "Worse than 1866 cholera." Stiff-upper-lipped Brits who had survived the Nazi air raids of World War II and the post-war rationing of food and other essentials must have thought they had descended another level in Dante's inferno. People gathered on the steps in front of St. Paul's Cathedral and prayed for salvation, convinced that this was the end of the world.

Bad government policy played a part in London's disaster. In 1307, King Edward I had decreed that coal should no longer be burned, as it was harming his subjects' "bodily health."

Unfortunately the decree was never respected, and more than six centuries later, coal was still widely burned in the U.K., both to generate electricity and to warm homes. To make matters worse, Britain was exporting its more valuable high-grade coal and using its dirty sulphur-laden coal at home.

Air pollution crises like London's occurred in Belgium's Meuse Valley; Poza Rica, Mexico; New York City; and Donora, Pennsylvania. The Donora disaster of 1948 was caused by air pollution from a steel mill combined with a temperature inversion that prevented the toxic emissions from dispersing. American Steel maintained that the disaster was an "act of God." The corporation ruthlessly defended itself against lawsuits from victims seeking compensation and never did admit responsibility.

The silver lining of the catastrophic air pollution experienced in Europe and North America during the 1950s is the legacy of laws passed to address the issue. The London smog led to the first clean air legislation in the world, while the American air quality crises generated public demand for the pioneering Clean Air Act of 1970. These laws set national standards for air quality; forced motor vehicle manufacturers to install filters, catalytic converters, and other pollution abatement technologies; and imposed standards on industrial sources of air pollution. The late Dr. David Bates worked at St. Bartholomew's Hospital in London during that fateful week in 1952 before moving to Canada and becoming one of the world's leading experts on the adverse health effects of air pollution. Dr. Bates wrote that "the London Disaster of 1952 should be commemorated; the many efforts to limit ambient air pollution that have occurred in the past 50 years are the proper memorial to those who were its unheralded victims."

Clean air laws have been one of the most remarkable successes in the history of modern government, saving millions of lives and generating trillions of dollars in net health and environmental benefits. It is estimated that the first 20 years of implementation of the U.S. Clean Air Act (from 1970 to 1990) imposed $523 billion in costs on businesses, consumers, and governments. However, the Clean Air Act also resulted in between $5.6 and $48.9 trillion in benefits during this period, with a central estimate of $22.2 trillion. The U.S. Environmental Protection Agency estimates that the full implementation of the Clean Air Act and various amendments to that law will prevent 230,000 premature deaths per year by 2020, with all health and environmental benefits producing an economic value that will exceed $2 trillion annually. The costs of achieving these benefits are a relatively paltry $65 billion per year. The EPA notes that under every scenario, the benefits of air quality regulation exceed the costs, with the most conservative estimate of benefits exceeding costs by a factor of three to one, the moderate estimate of benefits exceeding costs by 30 to one, and the high estimate of benefits exceeding costs by 90 to one.

You can walk onto a new car lot today and any vehicle you choose for a test drive will produce 90–100 percent fewer emissions of particulate matter, sulphur dioxide, and nitrogen oxides than the models sold in the early 1970s. That's directly a result of stringent legal standards, passed when health and environmental activists managed to persuade governments to pass legislation over the objections of their corporate opponents. Similarly, go to any gas station to purchase fuel, and the sulphur content will be at least 90 percent lower than it was 25 years ago. Again, progress produced by strong regulations. When Canada enacted a regulation in 2002 reducing the legal limit of sulphur in gasoline from 300 parts per million to 30, experts predicted that over the next 20 years, the regulation would prevent 11 million new cases of croup and pneumonia, five million restricted activity days for individuals suffering from asthma, 100,000 new cases of bronchitis, 9,000 emergency room or hospital admissions, and 2,000 premature deaths.

Graphs depicting ambient levels of various air pollutants in Canada, the U.S., and Western Europe generally follow the same pattern, rising steadily through the '50s, '60s, and '70s before levelling off and then plunging downwards. Levels of particulate matter in London today are only 1 percent of the peak levels reached in 1952. Between 1970 and 2012, emissions of the six main air pollutants—particles, ozone, lead, carbon monoxide, nitrogen oxides, and sulphur dioxide—plunged an average of 72 percent in the United States. Ambient levels of sulphur dioxide have dropped 80 percent since 1980. Lead concentrations in the air have fallen by 99 percent since 1980. All 41 areas of the U.S. that endured unhealthy levels of carbon monoxide in 1991 now meet the health-based national standard.

The pattern is the same in Canada. Since 1974, concentrations of carbon monoxide are down 90 percent and of sulfur dioxide are down 80 percent. Nitrogen dioxide concentrations are down 80 percent since monitoring began in 1980. Emissions of particulate matter and volatile organic compounds are down 30–40 percent. That's a lot of pollution not entering the atmosphere and thus not harming our respiratory or cardiovascular systems.

The health benefits of cleaner air are enormous and include fewer premature deaths, a decrease in the onset and exacerbation of asthma, improved development of children's lungs, fewer cases of lung cancer, and fewer heart attacks. Wait a minute, you might say, heart attacks? What does air pollution have to do with cardiovascular disease? Dr. Francois Reeves, a heart surgeon from Montreal and author of the book *Planet Heart: How an Unhealthy Environment Leads to Heart Disease* can answer that question. On a daily basis, Dr. Reeves performs catheterizations, angioplasties (mechanically widening obstructed or narrowed arteries), and heart bypass procedures (creating a new route for blood and oxygen to reach a patient's heart). He says, "I can tell how busy my day will be based on the air quality rating." Yet Reeves points out that when he was in medical school, students were taught that there were five main risk factors for heart disease: smoking, diabetes, high cholesterol, high blood pressure, and inherited genes. Air pollution wasn't even mentioned.

Scientists have only recently determined that inhaling polluted air affects our heart in much the same way as smoking, by damaging the arteries. Medical researchers and organizations such as the American Heart Association now agree that exposure to ambient particulate matter is an important risk factor for cardiovascular disease and also aggravates the classic risk factors. Air pollution can contribute to chest pain, abnormal heart rhythm, heart attacks, and congestive heart failure. Dr. Reeves identifies the absence of air pollution as one of the reasons why heart disease is almost non-existent in remote indigenous communities, such as the Tsimane people of Bolivia's Amazon rainforest. As Reeves concludes, air pollution "must now be added to the classic risk factors for heart disease."

European scientists have published information making it immediately clear how improvements in air quality provide tangible benefits for each and every one of us. Since the adverse health effects of exposure to particulate matter are partially reversible, in places that reduce particle pollution to the levels recommended by the World Health Organization, adults over 30 years old would live an average of 22 months longer. In other words, your life expectancy could jump by almost two full years as a result of improved air quality.

LET'S RETURN TO LOS ANGELES, 60 years after the nightmarish smog episodes of the 1950s. The concentration of volatile organic compounds (VOCs), an important component in the creation of smog, has dropped by approximately 98 percent in L.A. over the last half-century, from 100 parts per billion to 2 ppb. Cars are smaller and on most days you can see for miles. The iconic Hollywood sign is now routinely visible from downtown. At the turn of the 21st century, in L.A. and throughout the U.S., the air was far cleaner than it had been decades earlier. Medical researchers tracking the health of children's lungs in L.A. over the past 20 years have found encouraging improvements in lung function and development.

However, even in 2000, neighbourhoods in Riverside, California, an hour's drive east of Los Angeles, still endured some of the dirtiest air in the U.S. Riverside has a large volume of vehicle traffic and also endures air pollution funnelled eastward from L.A. On one in three days, air quality in Riverside fell well below the standards mandated by the Clean Air Act. By law, American cities' annual average could not exceed 15 micrograms of fine particles per cubic metre of air. People in Riverside were breathing almost twice as many microscopic particles of pollution as experts and the law deemed healthy.

Yet by 2014, for the first time, people in Riverside breathed air that met the annual health standard for fine particles. The average concentration of fine particle pollution in Riverside's air was cut in half since the 1990s. Considered inconceivable just a decade ago, achieving the federal air quality target in Southern California and nationwide is "perhaps one of the nation's greatest environmental success stories," according to Sam Atwood of the Los Angeles Basin's air quality agency.

Because new scientific evidence demonstrates that exposure to particle pollution damages human health at lower levels than previously believed, beginning in 2015 the EPA tightened the national standard even further, lowering it from 15 micrograms per cubic metre to 12 micrograms. "For health effects, the big one now is premature mortality," said Scott Fruin, an assistant professor at the University of Southern California who studies the health impacts of fine particulate matter ($PM_{2.5}$). "We see reduced life expectancy and higher chances of developing cardiovascular disease in places where the standard isn't met." Riverside and dozens of other American communities across the U.S. are again out of compliance because of the new standard. However, based on the success of their previous efforts, they anticipate meeting it by 2020. "The new standard will have enormous benefits . . . so many lives saved," said Janice Nolen, assistant vice-president of national policy at the American Lung Association. "If you've never had air that clean, you have no idea how big the benefits will be yet, and we're excited about that."

Riverside isn't an isolated example but rather the last domino to fall in a nationwide trend. Over the past ten years, the amount of $PM_{2.5}$ emitted in the U.S. has declined 45 percent according to the EPA. Much of the credit goes to cleaner diesel engines, mandated by stringent national standards. It would take an astonishing 60 new diesel trucks to produce the same amount of PM2.5 as a truck manufactured in 1988. Big industrial polluters have also made strides. Over the last ten years, U.S. industries have reduced total emissions of fine particulates by about 500,000 annual tons, or 26 percent.

In the early 1990s, I was a student at McGill University in Montreal. I ran on the school's cross-country team and was also training for the Boston Marathon. On my three- to four-hour runs, I would return to the locker room with nostrils full of soot. Blowing my nose would darken a Kleenex. My nostrils had done their job, filtering out some of the particulate matter that would otherwise have entered my lungs and potentially harmed my cardiovascular system. At the time it didn't bother me. Now I look back with a certain degree of dismay at my youthful ignorance. Microscopic particles of pollution almost certainly entered my lungs and bloodstream. But given recent improvements in Montreal's air quality, it is unlikely that a long run around the city today would produce a nose full of black particles.

DESPITE THE REMARKABLE PROGRESS in improving air quality over the past 45 years, air pollution continues to inflict a substantial toll even in Canada, the U.S., and Europe. The Canadian Medical Association estimates that air pollution causes over 20,000 premature deaths annually. In the U.S., about 200,000 premature deaths are linked to air pollution. The European Environment Agency estimates 400,000 premature deaths each year. In the world's wealthy nations, there is more work to be done, particularly on behalf of communities that are poor, socially marginalized, or otherwise vulnerable. The burden of pollution is disproportionately borne by these people.

In 2008, my wife, Margot (who is also an environmental lawyer), and I went on a "toxic tour" of Sarnia, Ontario, known as Canada's Chemical Valley. The things environmental lawyers do on their holidays! Our host was a courageous fellow named Ron Plain, a member of the Aamjiwnaang First Nation and campaigner for a group called Environmental Defence. Ron's ancestors have lived in this area for thousands of years.

The Aamjiwnaang reserve is surrounded by dozens of oil refineries and chemical plants. More than 60 large industrial facilities are located within a 25-kilometre (15-mile) radius of the community, pumping out hundreds of thousands of tonnes of toxic chemicals annually. It was shocking to drive around Sarnia, stopping at one massive chemical complex after another. There were hundreds of smokestacks belching pollutants into the air. We parked briefly outside a vinyl chloride plant where we all felt like gagging, the chemical stench was so powerful. We walked beside a small creek where a sign warned, "Keep Out. Talfourd Creek contains toxic substances known to cause serious health risks." We stopped for lunch at the Leaky Tank Truckstop, where I couldn't eat my grilled cheese sandwich and fries because my stomach was tied in knots. We marvelled at the twisted sense of humour behind the name of the restaurant and the wall-sized mural of an 18-wheeler leaking fluorescent toxic chemicals onto the highway. We stared in disbelief at the former location of the Aamjiwnaang daycare centre, right across the highway from several chemical factories. We ended the tour by walking quietly around an Aboriginal cemetery fenced in on three sides by refineries and petrochemical plants.

I've never smelled fouler air than I did that day in Sarnia. My body rebelled. I've always been blessed with robust physical health, allowing me to complete marathons, ultramarathons, and Ironman triathlons. In my whole life, I've only had two or three headaches. But after a couple of hours on the Toxic Tour, my eyes felt as though they'd been rubbed with sandpaper. My throat was raw and my head throbbed with a debilitating headache. In time those symptoms eased, but to this day my heart is sickened by what I saw. In one of the wealthiest, best educated, most technologically advanced nations in the world, with an international reputation for being socially progressive, was one of the most disgraceful displays of human indifference that I have ever seen. If a couple of hours in the poisoned atmosphere of Chemical Valley was enough to make me feel ill, what kind of health problems must be endured by the people who live there?

Our comfortable lifestyle carries a prodigious cost, paid by people who live in places like Aamjiwnaang and Sarnia. The World Health Organization confirmed that Sarnia has the worst air quality in Canada, so it wasn't my imagination or an unusual day. My Toxic Tour experience brought home to me that it's not enough to take action just to protect ourselves and our families from environmental hazards. We have an obligation to ensure that governments and industries take the necessary steps to ensure that everyone is adequately protected. Places like Chemical Valley violate people's fundamental right to live in a healthy environment. They are a stain on our collective conscience that needs to be scrubbed clean.

IN THE WORLD'S LESS WEALTHY COUNTRIES, air pollution inflicts even more deadly and devastating damage on human health. The

World Health Organization estimates that seven million premature deaths every year are caused by air pollution. Roughly half of these deaths are from poor indoor air quality caused by cooking and heating with wood, dung, or low quality coal. The other 3.5 million deaths are caused by outdoor air pollution. These premature deaths, as well as millions of illnesses, are not inevitable and could be prevented with stronger environmental laws and policies.

In 2013, Harbin, China, experienced a smog episode similar to London's of 1952, with PM2.5 levels 40 times higher than World Health Organization guidelines. Persistent smog and other ecological disasters are fuelling public dissatisfaction and anger in China. Antipollution demonstrations are increasingly common. Chinese leaders must be aware of the risk that protests related to environmental pollution pose to the political stability of a country where people are quickly gaining wealth but want a better quality of life. Environmental protests were a lynchpin in the pro-democracy movement in Eastern Europe in the 1980s, but thought to be insignificant by the Soviet authorities who were out of touch with the emerging zeitgeist.

In recognition of the magnitude and urgency of the air pollution crisis gripping their country, Chinese leaders have pledged to spend more than $300 billion over the next five years to improve air quality. China is also setting, and in some cases surpassing, ambitious targets for clean renewable energy, high-speed rail, public transit, electric vehicles, and recharging infrastructure.

In 2012, I was invited to Mexico City to provide a keynote presentation at a conference on human rights and the environment. Despite some trepidation about the city's notorious air quality, I packed my running gear. Approaching Mexico City at night in an airplane was a mind-blowing experience. After first spotting the metropolis out the window, I was astonished as our approach went on and on. For many minutes, the city's constellations of light twinkled beneath us, as though the night sky had been flipped upside down. I finally grasped the mega-city's size.

While there, I went for several morning runs in Chapultepec Park in the middle of the city. I was pleasantly surprised by how clean the air seemed. My lungs still gasped for oxygen, but the elevation (over 2,200 metres) was the culprit, not the pollution. Air quality in Mexico City, once among the worst in the world, is still not ideal, but has improved substantially in recent years, with reductions exceeding 50 percent for most pollutants. Levels of lead in the air are down over 90 percent since 1990. Ozone levels are down 75 percent. Particles are down 70 percent. The government has forced factories to clean up, close, or, sub-optimally, relocate to areas with less air pollution. Fuels have been reformulated to reduce pollution. Rules limit the use of cars, while public transit systems have been expanded. Without these actions, Mexico City "would be a living hell" according to Exequiel Ezcurra, former head of Mexico's National Institute of Ecology. With 20 million inhabitants and a growing number of cars, there are still major obstacles to securing clean air in the Mexican capital. Nobel Prize winning scientist Mario Molina said, "If the government decides to do something about it, it can be done. There's really no excuse not to do more."

WE KNOW, BASED ON the experiences of the past 50 years, that by enacting and enforcing progressively stronger environmental laws and policies we can reduce air pollution's terrible toll. But too often old-school industries will twist the scientific evidence, falsely assert that the costs outweigh the benefits, and predict economic disaster if the rules are tightened. In 2015, this tired debate is playing out again in the U.S. in response to President Obama's new regulations on the coal industry and a proposal to strengthen the rules on ozone pollution. It is playing out in Canada as the oil and gas industry persuades governments to drag their heels in regulating their emissions. Although health experts agree that these rules are overdue and eminently justified by cost-benefit analysis, industry on both sides of the border is mounting a massive legal, political, and public relations battle in an attempt to defeat them. The public, health care advocates, environmentalists, and medical professionals need to help stiffen the spine of politicians to ensure that they do what is right. Then we can all breathe a little easier. Thanks to cleaner air, hundreds of millions of people already live happier, longer, more productive lives.

"Pessimists are usually right and optimists are usually wrong, but all the great changes have been accomplished by optimists."
—Thomas L. Friedman

Chapter 5
Saving the Ozone Layer

READING DAVID SUZUKI'S BOOK *It's a Matter of Survival* in 1992 made me want to become an environmental lawyer. I still have the book, its cover faded, pages bristling with yellow sticky notes. The book was based on an eight-part CBC Radio series that prompted almost 20,000 people to write Suzuki. Real paper-and-ink letters, in the era before email. In the opening chapter, Suzuki painted a harrowing picture of what the world would look like in 2040 unless immediate action was taken to address the ongoing destruction of the Earth's protective ozone layer. Based on the best scientific evidence available at that time, Suzuki forecast devastating impacts on food production, decimation of phytoplankton and zooplankton (the base of the marine food chain and major producers of oxygen), and spikes in infectious diseases, cataracts, and skin cancer, both benign and malignant. He quoted a scientist who speculated that ozone depletion would force Australians to morph from outdoor enthusiasts into people who lived largely indoors. In passing, Suzuki referred to the unprecedented development of international laws to protect the ozone layer. It was belatedly apparent to me that my law degree, which was gathering dust, could be a valuable tool in doing my part to protect our beleaguered planet.

Only 25 years later, one of the most remarkable environmental success stories in human history is largely forgotten. The tale of how humans inadvertently imperiled and then rapidly saved the ozone layer serves as a tremendous example of both humanity's ability to jeopardize the future of life on this planet and the power of nations working together to solve a common problem.

The word ozone comes from the Greek term for smell. The distinctive sulphurous smell of ozone, associated with lightning, was first mentioned by Homer in *The Iliad* and *The Odyssey* in 850 BC. The first scientist to identify ozone as a distinct substance was Christian Friedrich Schönbein, a German-Swiss chemist famous for inventing the hydrogen fuel cell at the same time as Welsh physicist William Grove. While doing experiments with water in the 1830s, Schönbein noticed a peculiar smell in his lab—ozone—similar to that occurring in a thunderstorm. Ozone is formed by chemical reactions involving oxygen molecules. Sunlight breaks oxygen molecules (O_2) into two oxygen atoms (O_1), and each of these liberated atoms binds with another oxygen molecule to form ozone (O_3).

The presence of the ozone layer was discovered by French physicists Charles Fabry and Henri Buisson in 1913, but it was British meteorologist Charles Dobson who first explained the importance of stratospheric ozone to life on Earth. The ozone layer, 10 to 30 kilometres above the planet's surface, acts like sunscreen, filtering out harmful ultraviolet (UV) radiation from the sun. UV radiation can damage plants, cause cataracts, suppress the immune system, damage DNA, and trigger cancer. In a potentially confusing twist, ozone close to the Earth's surface is a harmful air pollutant that damages crops, trees, and other plants and can adversely affect the human respiratory system. However, without the thin, invisible shield provided by the stratospheric ozone layer, life on Earth would be impossible. In the words of NASA scientist Pawan Bhartia, "The evolution of life is tied to the evolution of ozone. It is critical for the longtime survival of our species."

In a seemingly unrelated development, a new group of chemicals known as chlorofluorochemicals (CFCs) were synthesized by Belgian scientist Frédéric Swarts in the 1890s. CFCs were regarded as potentially valuable because they were chemically inert, meaning they did not react with other substances. The process of synthesizing CFCs was improved by American chemist Thomas Midgley, Jr., in the late 1920s, reducing their cost. At a meeting of the American Chemical Society, Midgley demonstrated the non-toxic and non-flammable nature of CFCs by inhaling the gas and blowing out a candle.

Midgley was also involved in developing a lead additive for gasoline, a product that harmed the intellectual development of hundreds of millions of children over a period of decades. He somehow managed to play a key role in two of *Time* magazine's 50 worst inventions of all time, leading writer Bill Bryson to observe that Midgley possessed "an instinct for the regrettable that was almost uncanny."

Because CFCs were cheap, non-reactive, non-flammable, and seemingly non-toxic, they became widely used: in coolants used in refrigerators, freezers, and air conditioners; as propellants in consumer products ranging from fire extinguishers and asthma inhalers to hairspray and deodorant; as foam-blowing agents to make myriad Styrofoam products such as disposable coffee cups; and as solvents.

In 1957, British scientists conducting research in the Antarctic began monitoring the ozone layer, using a gizmo called a Dobson spectrophotometer (invented by Charles Dobson) to measure the amount of ozone between the earth's surface and outer space. It turns out that if all of the ozone molecules in the stratosphere were put together and spread equally across the globe, they would form a layer only 3 millimetres thick, about the thickness of two pennies.

In 1969, another remarkable human invention took flight. Supersonic airplanes, such as the Concorde, were designed to fly faster than the speed of sound, enabling transatlantic flights from New York to Paris in less than three and a half hours. Two American chemistry professors, Halstead Harrison and Harold Johnston, unleashed a firestorm of controversy by speculating that nitrogen oxide emissions from the exhaust of these airplanes could inflict serious damage on the ozone layer. The U.S. government took the allegations seriously and formed a team of scientific experts to conduct research into this question. Studies confirmed that nitrogen oxides would harm the ozone layer but the limited commercial success of supersonic aircraft resulted in negligible impacts. Similar concerns about emissions from the nascent space shuttle program resulted in the same conclusion.

On a voyage from England to the Antarctic aboard the research ship RRS *Shackleton* in 1971, British scientist James Lovelock took 50 air samples at regular intervals over a distance of thousands of kilometres and was astonished to find a new manmade molecule, chlorofluorocarbon, present in all of them. The concentrations were low—measured in the parts per trillion—and based on his understanding of the inert nature of CFCs, Lovelock concluded that there was no problem.

Lovelock's discovery triggered the curiosity of Sherwood Rowland, a chemistry professor at the University of California (Irvine). According to Rowland, "Here was a new compound that had not been in the Earth's atmosphere before, and the question was: What will happen to it?" Rowland was delighted that a bright young student from Mexico named Mario Molina joined him in 1973 to conduct postdoctoral research. Faced with numerous projects, Molina picked CFCs because "to me at that time, it was really a fascinating question." The two scientists began collaborating on a project examining the atmospheric fate of CFCs.

Molina and Rowland hypothesized that after being released into the environment, CFCs were rising up to the stratosphere and participating in a chain of chemical reactions that released chlorine atoms. The liberated chlorine atoms then reacted with ozone molecules to disrupt the natural equilibrium of ozone formation and destruction, effectively thinning the ozone layer and decreasing its protective capacity. In June 1974, these two hitherto unknown scientists published a two-page article in the highly regarded journal *Nature*, warning that the use of CFCs was likely damaging Earth's stratospheric ozone layer. In typically guarded scientific language, the researchers concluded, "It seems quite clear that the atmosphere has only a finite capacity for absorbing chlorine atoms produced in the stratosphere, and that important consequences may result." They estimated that a chain of chemical reactions resulted in each chlorine atom being capable of breaking down as many as 100,000 ozone molecules. To make matters worse, once released into the environment, CFCs were expected to stay in the atmosphere for 75 to 100 years.

The response to their paper was muted. As Molina said in hindsight, this should not have been a surprise: "We were talking about this invisible gas rising in the atmosphere to affect an invisible layer that was protecting us from invisible rays." To elevate the issue's profile, later in 1974 Rowland and Molina held a press conference at a meeting of the American Chemical Society and called for a complete ban on releases of CFCs into the atmosphere. Rowland stated, "The original question of chiefly scientific interest has now been converted into a very significant global environmental problem—the depletion of stratospheric ozone by the chlorine contained in the chlorofluorocarbons." He added that his wife found 15 spray cans in their home and threw them out, leading him to think, "15 down and 6 billion to go." A *Globe and Mail* headline on August 21, 1974, asked, "Is handy aerosol can a life threat? Ozone layer perhaps in danger." American newspaper headlines warned "Death to Ozone" and "Aerosol spray cans may hold doomsday threat."

At the time of the Rowland and Molina report, it was estimated that global releases of CFCs were about one billion pounds per year. It was a multibillion dollar industry employing tens of thousands of people. Not surprisingly, the industry reacted strongly to the ozone depletion hypothesis. The two scientists were pilloried by industry spokespersons who claimed it was absurd to suggest that ordinary household items could be having a negative effect on the global atmosphere. DuPont ran full-page newspaper ads defending CFCs. According to industry, there was "no concrete evidence" that their products were causing a problem. British meteorologist Richard Scorer, editor of the *International Journal of Air Pollution*, was hired by the chemical industry to attack Rowland and Molina's credibility. Scorer barnstormed across the U.S. on DuPont's dime, stating that the ozone depletion theory was "a science fiction tale," "a scare story," "a load of rubbish," and "utter nonsense." In 1974, Canadian scientist Harold Schiff, a chemist at York University, acknowledged that "there are many ifs and unknowns" in terms of our understanding of the stratospheric ozone layer.

In 1976, the U.S. National Academy of Sciences produced a consensus report concluding that Molina and Rowland had got the science right in their description of chemical reactions in the stratosphere involving CFCs and ozone. Soon afterwards, a few governments and corporations began to take baby steps to tackle the problem. Responding to scientific evidence and public pressure, the U.S., Canada, Sweden, and Norway banned the use of CFCs in a handful of consumer items including hairspray and deodorant. SC Johnson, a major American manufacturer of household cleaning products, accepted the science and announced it would no longer use CFCs. However, the majority of CFC applications were allowed to continue unabated while further scientific research was undertaken.

In spring 1985, Joseph Farman and a team of British scientists revealed a sharp and dramatic decline in the thickness of the ozone layer over the Antarctic. This was the first tangible evidence that Rowland and Molina's hypothesis was correct. Although what was described was technically a thinning of ozone, the media described it with more apocalyptic imagery as a hole in the ozone layer. Farman and his colleagues actually made their discovery in 1981, but their readings were so low that the scientists believed their instruments had malfunctioned.

The impact of the British research was initially blunted by the fact that they only had data from one place. A few months later, scientists from NASA and the U.S. National Oceanic and Atmospheric Agency put together additional pieces of information from other locations and reached the same conclusion—the ozone layer was rapidly thinning. NASA satellites should have identified the problem earlier, but the results had been dismissed by a computer program as "bad data" or errors and filtered out because, again, the readings were so low. It was not bad data or instrument error. It was a real phenomenon that exceeded the worst-case scenarios of scientists and shocked the public. As NASA's Pawan Bhartia said, "After that, all hell broke loose." Influential American Senator John Chafee acknowledged, "There is a very real possibility that man—through ignorance or indifference or both—has irreversibly altered the ability of our atmosphere to support life."

International negotiations that had been dragging on for years suddenly accelerated, leading to a framework treaty in 1985 called the Vienna Convention on Substances that Deplete the Ozone Layer. The Vienna Convention was a pledge by governments to address the issue, but it was short on details and had no specific targets or timelines for action. While some progressive businesses phased out the use of CFCs, widespread chemical industry opposition continued and found sympathetic ears in some governments. President Reagan's Interior Secretary Donald Hodel proposed dropping the international treaty approach in favour of a Personal Protection Plan focused on encouraging Americans to wear hats and sunglasses. The "Rayban Plan" was widely ridiculed and abandoned, and the U.S. ended up playing a constructive role in the ongoing treaty negotiations.

Another hammer dropped in 1987. Scientists discovered a second "hole" in the ozone layer over the Arctic, sparking front-page headlines and a sense that the world was facing its first global environmental emergency. John Gille of the National Center for Atmospheric Research concluded, "We've found more than the smoking gun. We've found the corpse." Dramatic computer-generated images of the ozone "hole" galvanized public opinion, which was counter-balanced by fierce lobbying and stonewalling by industry.

The chemical industry's strategy on the ozone issue was copied directly from the tobacco industry playbook. First, question and dispute the evidence, including ad hominem attacks on leading scientists. DuPont chair Richard E. Heckert claimed, "Scientific evidence does not point to the need for dramatic CFC emission reductions." Until 1980, DuPont continued to insist that "no ozone depletion has ever been detected." Or, if thinning of the stratospheric ozone layer was occurring, industry blamed volcanoes, natural variation, atmospheric dynamics driven by extremely cold temperatures in the polar regions, or the 11-year solar cycle. CFC manufacturers hired Hill+Knowlton, one of the world's largest public relations firms, to fight against potential government regulation. They hired Fred Singer, one of a handful of scientists who, in the face of overwhelming evidence, steadfastly defended corporate interests including tobacco, asbestos, CFCs, and fossil fuels.

According to Singer and his ilk, smoking didn't cause lung cancer, asbestos didn't cause mesothelioma, and CFCs didn't harm the ozone layer. British scientist Richard Scorer derided Molina and Rowland as "doomsayers." The arch-conservative Cato Institute alleged that NASA's warnings about a potential ozone hole in the northern hemisphere "were exquisitely timed to bolster the agency's budget requests."

The second element of the industry strategy was to deny the existence or availability of any alternatives. Potential substitutes were said to be toxic, flammable, and expensive. Third, the CFC industry predicted dire economic consequences and ignored all potential benefits of regulation. The industry-sponsored Alliance for Responsible Atmospheric Policy argued that phasing out CFCs "would result in devastating impacts on the quality of modern lifestyles as well as on national economies." The Fraser Institute, a right-wing Canadian think tank, warned that cutting back on the use of CFCs "would require large sacrifices on the part of everyone." Banning CFCs allegedly would cause 20 to 40 million deaths annually because of the "collapse of refrigeration." According to the Fraser Institute, "More people would die from food poisoning as a consequence of inadequate refrigeration than would die from depleting ozone." Five million children would die unnecessarily because refrigerated vaccines would not be available for immunization.

To their credit, governments acted swiftly despite industry opposition, negotiating a series of specific commitments to phase out the use of CFCs and other ozone-depleting chemicals. The Montreal Protocol of 1987 was the breakthrough agreement with 24 countries pledging to cut CFC emissions in half by the end of the century. By 1989, the Antarctic ozone "hole" was two-and-a-half times larger than Canada. As the scientific evidence evolved rapidly, so did the international negotiations. Subsequent treaty talks in London, Copenhagen, Vienna, and Beijing accelerated the phase-out schedule and expanded the agreement to cover a broader range of chemicals. In recognition of the fact that the rich countries had created the lion's share of the problem, parties to the Montreal Protocol set up a billion-dollar fund to assist developing nations in the shift to safer substitutes. Around the world, comprehensive laws were passed to regulate all stages of the life cycle of products containing or manufactured with these chemicals. The U.S. even established an Ozone-Depleting Chemicals Tax, which raised over $3 billion for the development of alternatives.

It helped smooth the process when industrial giants like DuPont finally unveiled effective and profitable substitutes for ozone-depleting substances. Sherwood Rowland and Mario Molina won the Nobel Prize in Chemistry in 1995 for their scientific breakthrough, along with Dutch scientist Paul Crutzen, who had shown that nitrogen oxides also accelerate ozone depletion. In his Nobel acceptance speech, Molina sent a message to scientists around the world stating, "We should speak out when we feel it's important for society to change." Similarly Rowland said, "If you believe that you have found something that can affect the environment, isn't it your responsibility to do something about it, enough so that action actually takes place?"

Many thought that the promises made in the Montreal Protocol would never be kept. It was cited as a textbook case of the free rider problem from economics, where every country would benefit from the protection of the ozone layer, even if they didn't contribute to the solution. Experts predicted that the agreement would eventually fall apart. The opposite happened. Countries that had signed on acted. More countries joined. They also acted. Commitments were kept, strengthened, and kept again. Richard Benedick, the chief U.S. negotiator, called the treaty "unique in the annals of international diplomacy" and a "forerunner of an evolving global diplomacy, through which nations accept common responsibility for stewardship of the planet."

The Montreal Protocol is now widely hailed as the most successful international environmental treaty ever negotiated. Every country in the world is a party to it. The protocol controls the use of approximately 100 ozone-depleting chemicals including CFCs, hydrochlorofluorocarbons (HCFCs), hydrofluorocarbons (HFCs), halons, hydrobromofluorocarbons, chlorobromomethane, methyl chloroform, and methyl bromide. CFCs were completely phased out by 2010, while deadlines for ending the use of methyl bromide, HCFCs, and HFCs stretch as late as 2040 for some developing countries. Globally, CFC production peaked in 1988 and has since fallen by 99 percent. As a result, in most regions of the world, thinning of the ozone layer stopped by 1998, and full recovery is expected to occur in the mid 21st century.

As an added bonus, many of the new technologies developed to protect the ozone layer also save energy and reduce greenhouse gas emissions. A comprehensive review of the American efforts to reduce ozone-depleting substances concluded: "Contrary to early predictions, meeting this protocol goal did not severely disrupt the U.S. economy, trigger massive job losses, or deny popular products or services to consumers. With considerable ingenuity, and aggressive investment in innovation, many U.S. industries eliminated

CFC use more quickly, at lower cost, or with greater environmental benefits than observers once predicted." The switch from CFC-based propellants to substitutes reduced industry costs by as much as 80 percent, saving U.S. businesses and consumers over $1 billion. Phasing out ozone-depleting chemicals also reduced greenhouse gas emissions equivalent to saving 1.2 trillion gallons of gas, enough for almost five billion round-trip journeys from New York to Los Angeles. Analyses by Environment Canada and the U.S. EPA show that benefits outweighed costs by a ten-to-one ratio, not even including the human health benefits.

Just a generation ago, many of the products in the average home were produced with or contained ozone-depleting substances. Today, refrigerators, vehicles, computers, furniture, fire extinguishers, hairspray, foam packaging, and air conditioners are made in ways that do not jeopardize the ozone layer. Some unfinished business remains. Ozone-depleting substances persist in the atmosphere for decades, so the past use of these chemicals continues to slow the recovery process. There is a black market in CFCs and other illegal ozone-depleting chemicals. Some of the substitutes are not entirely benign, as they are powerful greenhouse gases. Schedules for eliminating HCFCs and methyl bromide should be accelerated. There are potential complications from climate change that could delay the recovery of the ozone layer. However, in comparison to the catastrophe that humanity avoided, these are relatively niggling concerns. A recent study estimated that if the use of CFCs had continued unabated, then in 2065 you could have received a serious sunburn in a city like Washington, D.C., in just five minutes. As much of half of the world's agricultural production could have been wiped out. The actions taken pursuant to the Montreal Protocol prevented millions of people from suffering cataracts, skin cancer, and compromised immune systems, providing humankind with trillions of dollars in net benefits. By 2165, in the U.S. alone, actions to protect the ozone layer will have prevented 6.3 million deaths from skin cancer and produced over $4 trillion in health benefits.

David Suzuki wrote that the discovery of holes in the ozone "underscored our fundamental ignorance of how our actions affect atmospheric chemistry." The ozone crisis also demonstrated, in unprecedented fashion, that humans had evolved from a large-brained hunter-gatherer whose environmental impacts were highly localized into a species with the capacity to unintentionally wreak havoc at a global scale.

Satellite images from 2014 showed that the ozone "holes" over the Arctic and Antarctic regions have shrunk to their smallest size in more than a decade. This is enormously good news. Mike Newchurch of the Global Hydrology and Climate Center stated, "Up there, in the stratosphere, the Montreal Protocol seems to be working." Humanity dodged a bullet through strong environmental laws informed by the best available science, international cooperation, technological innovation, modified behaviour, and the application of the precautionary principle. As we grapple with the ongoing challenges of climate change and biodiversity loss, it is reassuring to know that we have come to terms with global environmental problems in the past and managed to avoid prognostications of planetary catastrophe through swift and effective action.

I was nine years old when Rowland and Molina issued their initial warning and cannot recall hearing anything about it at school or at home. There was no mention of their discovery in our local paper, the *Calgary Herald*, in 1974. By the time Canada banned CFCs in hairspray and other consumer items, I was a geeky, fair-skinned teenager. Oblivious to the existence of the ozone layer or its existential threat and desperate to acquire the beautiful tanned skin of the cool kids at school, I would lie on a towel in our Calgary backyard in my gym shorts reading piles of books. My body glistened with Hawaiian Tropic tanning oil, whose exotic and addictive coconut scent I can still vividly recall. Sun protection factor (SPF): zero. My initial sunburn would be vicious but eventually evolved into a fairly respectable tan. It seems crazy now. I have no recollection of ever wearing sunscreen as a child, or even being asked to do so by my mom. At the time, there was no such thing as the UV index, a tool developed by Canadian scientists in 1992 and adopted by the World Health Organization in 1994. The UV index informs people about daily levels of UVB radiation and recommends specific protective actions.

When news of the thinning ozone layer broke in 1985, I had just returned from a lengthy trip to Spain with my brother, inspired by one of my boyhood heroes. Ernest Hemingway had immortalized the seven-day San Fermín festival (widely known as the running of the bulls) in his novel *The Sun Also Rises*. We drank copious quantities of red wine (often mixed by Basque youths with Coca-Cola to make a surprisingly palatable drink called *calimocho*), ran with the bulls at Pamplona, and spent a lot of time in the blazing Iberian sun. After two months in Spain, I was deeply bronzed, my arm and leg hairs had been bleached blond, and I was scarcely recognizable even to friends and family. Total sunscreen use: zero.

Today, my fair-skinned daughter Meredith can hardly step outside without being admonished to wear a hat and slather on sunscreen. We avoid the beach at midday and keep a wary eye on the UV index. It is sobering to read the words of Dr. Gary Sibbald, a former director of the Canadian Dermatology Association, who said, "The skin never forgets. All the ultraviolet rays you have been exposed to in your lifetime will add up." There is visible sun damage on my face, arms, and legs. Several years ago, I had a mole removed from my leg after my family doctor expressed concern about its growth. Lab tests showed no indication of skin cancer, but the threat remains. I shudder to think of what life would have been like for Meredith, her friends, and their children if not for the rapid scientific breakthroughs of the 1970s and 1980s and aggressive action taken by a generation of political leaders who listened to scientists and discounted the economic doomsday warnings of vested interests.

*"Optimism is the faith that leads to achievement.
Nothing can be done without hope and confidence."*
—Helen Keller

Chapter 6
Taps, Toilets, and Farms

HUMAN HEALTH AND WELL-BEING depend on access to safe drinking water, adequate sanitation infrastructure, and sufficient quantities of nutritious food. For centuries, contaminated water and food killed millions by exposing them to harmful micro-organisms including bacteria and viruses. Cholera, diarrhea, and dysentery were frequent maladies with sometimes fatal consequences. Similarly, starvation and malnutrition contributed to millions of premature deaths. The provision of access to safe drinking water and sanitation is one of the greatest human health achievements of all time. Enough food is produced annually to easily feed all of our over seven billion humans, if only it were equitably distributed. All of this progress, however, has inflicted enormous environmental costs. The challenge now is to continue making progress in fulfilling human needs while reducing levels of ecological harm.

TURN ON THE TAP. The overwhelming majority of individuals reading this book will be able to quaff a glass of nice cold water without a second thought about its quality or safety. Modern water treatment and distribution systems now offer more than four billion people the luxury of drinking clean water straight from a tap for less than a penny per glass. Access to safe drinking water is among the most important factors in the doubling of life expectancy over the past century and a half.

Despite medical recognition of the importance of clean water, industrialization led to countless contaminated and degraded water bodies by the late 1960s. Infamously, the Cuyahoga River in Ohio caught on fire in 1969. It boggles the mind to imagine a river burning, and yet levels of hydrocarbon pollution in the Cuyahoga were so high that the fire raged. This debacle, and others, helped spark the development of the U.S. Clean Water Act and the Safe Drinking Water Act, kin to the pioneering Clean Air Act described in Chapter 5. These laws have been emulated by many countries.

Vancouver, where Margot and I lived for many years, has some of the best tap water in the world. Prescient politicians protected mountainous watersheds on the city's north shore, so rain collects in reservoirs before being piped downhill for filtration, treatment, and distribution. In 2006, a vicious storm unleashed torrential downpours and winds exceeding 120 kilometres per hour, knocking down thousands of trees and triggering landslides that sent tonnes of sediment into the water supply. Tap water turned cloudy, even muddy. It wasn't pretty, but it wasn't necessarily unhealthy. However, micro-organisms that cause gastrointestinal illness can avoid chlorination by adhering to sediments, so as a precautionary measure, public health authorities advised residents to boil water for at least a minute prior to drinking, brushing teeth, or washing fruits and vegetables. Some people were indignant, others ignored the warnings, but most people complied. There was a run on bottled water that led to shortages and some confrontations. A few coffee shops closed temporarily. At worst, the boil water advisory was mildly inconvenient. There was no outbreak of waterborne illness. But for ten days, Vancouverites had a small taste of the plight that still faces hundreds of millions of people.

Imagine having to walk every day for 30 minutes, an hour, or even longer to fetch a container of water for your family's basic needs. This is a task that often falls on the shoulders of women and girls in Africa and South Asia. These are conditions that no person should be forced to endure. In the late 1990s, the United Nations established the Millennium Development Goals (MDGs), a series of ambitious targets intended to improve the lives of the poorest people in the world by 2015. Among the targets was reducing by half the number of people lacking access to safe drinking water and adequate sanitation. To the surprise of many, including critics who had derided the MDGs as mere posturing, the safe water goal was among the first to be achieved, ahead of schedule. Between 1990 and 2014, an amazing 2.3 billion people gained access to improved drinking water sources. That brings the total to 9 out of 10 people globally. In wealthy countries like Canada and the U.S., as well as 90 other countries, diseases like cholera are now almost unheard of, as at least 98 percent of the population has access to safe drinking water. On the other hand, it is important to remember that roughly 700 million people, the majority of whom live in sub-Saharan Africa, still lack this access.

In 2010, after long campaigns by human rights activists such as Canada's tireless Maude Barlow and Bolivia's Pablo Solón, the UN General Assembly passed a resolution recognizing the right to water, with 124 nations voting in favour, none against, and 41 nations abstaining. The resolution stated, "The right to safe and clean drinking water is a human right that is essential for the full enjoyment of life and all human rights." Canada and the U.S. were among the nations that abstained. Later in 2010, the UN Human Rights Council issued a similar resolution. Dozens of countries—from Belgium and Brazil to Uruguay and the Ukraine—now recognize the right to water in constitutions, national legislation, or policy.

Recognizing that people have a legal right to safe drinking water and sanitation can dramatically improve the situation of the most vulnerable and marginalized people. In the early 1990s, when Nelson Mandela and his colleagues were writing a new South African constitution for the post-apartheid era, prominent legal experts advised against including the right to water. Making promises that couldn't be kept, the experts warned, could undermine public respect for government and the constitution at a critical time in rebuilding a traumatized nation. Fortunately, Mandela had the courage to stick to his conviction, and he ignored the so-called experts. As a result, the South African constitution explicitly states that every person has the right to safe drinking water. Since 1996, South Africa has made safe drinking water a national priority, incorporating this right as a primary objective of new laws on water management and municipal services and investing heavily in new infrastructure. Thanks to these actions, more than ten million predominantly poor and black South Africans have secured access to safe drinking water. Nelson Mandela described these gains as "amongst the most important achievements of democracy in our country."

The Kalahari San (formerly referred to as Bushmen) used the UN's legal recognition of the right to water to secure a major legal victory against the government of Botswana. For many years, the San had used water from a borehole in a national wildlife refuge. When the government denied them access to their water supply, arguing that it was inappropriate in an area now set aside for wildlife, the San sued, asserting that this deprivation violated their human rights. Citing the UN resolution on the human right to water, Botswana's High Court ruled in favour of the San and restored access to their precious water supply. Similarly, in Argentina, courts have ordered governments to provide communities with potable water, construct drinking water treatment facilities, ensure medical treatment for individuals harmed by contaminated drinking water, and carry out environmental remediation of polluted watersheds. In Chacras de la Merced, a poor community whose drinking water was being contaminated by inadequate wastewater treatment in an upstream municipality, local residents filed a lawsuit asserting their constitutional right to water. The court agreed that their right had been violated and ordered the government to upgrade the wastewater treatment plant and, in the interim, provide a supply of clean water to the residents.

Canada and the United States are among a dwindling number of nations that do not recognize their citizens' right to water. Despite being water wealthy, there are marginalized communities in these countries that even today lack access to safe drinking water and adequate sanitation. Canada's federal government estimates that there are approximately 5,000 homes in First Nations communities (representing an estimated 20,000-plus residents) that lack running water and indoor toilets. The government admits that "the incidence of waterborne diseases is several times higher in First Nations communities than in the general population, in part because of the inadequate or non-existent water treatment systems." Many of these deplorable situations have been dragging on for years or even decades. In the U.S., more than 30,000 low-income people in Detroit had their water supply cut off in 2014 because poverty prevented them from paying their utility bills on time. These situations in Canada and the U.S. clearly violate the human right to water.

The right to water is not a silver bullet that will address all aspects of the world's water crisis. However, it is a powerful tool that can be used to focus attention and resources on improving access to water for those individuals and communities who currently endure the hardships imposed by the absence of safe water.

Another positive story involves bottled water sales, which once rose like a rocket year after year but are now falling in many Canadian cities. Tap water is almost always a superior choice both environmentally and economically. Hundreds of cities, responding to pressure from residents, have phased out or reduced spending on bottled water. Hundreds of restaurants have introduced similar policies, encouraging customers to drink free tap water instead of expensive bottled water. Vancouver residents exemplify the trend of choosing tap water more often, with bottled water purchases declining from 21 percent to 11 percent of water consumption in the last three years. This is part of a nationwide trend, reported by Statistics Canada, as bottled water consumption is on a steady decline from a peak reached in 2007. Bottled water sales have also fallen in a number of Western European countries (e.g., France,

Italy, Portugal) but in the U.S. and many developing countries, sales are still rising.

ONE OF THE MAJOR CAUSES of unsafe drinking water is fecal contamination, as seven billion people and their livestock generate an enormous volume of bacteria-laden excrement. I recall reading an interview with an American woman who, at the time, was the oldest living person in the U.S., born in the late 1800s. When asked what she considered to be the greatest invention of her long lifespan, she gave a surprising answer. It wasn't airplanes, space travel, or the Internet. The invention she believed had the greatest impact on her quality of life was the flush toilet. As a child, she and her family had used bedpans and an outhouse. She loathed emptying the bedpans, her chore every morning. Yet virtually everyone reading this book will take indoor plumbing for granted. Our body's wastes are flushed away without any thought to where the pipes will take them.

Toilets and sewage treatment systems improved people's quality of life, reduced the spread of disease, and protected aquatic ecosystems. But even in a country as developed as France, the majority of houses lacked indoor plumbing as recently as the 1950s. Many countries, including Canada, are in the process of upgrading sewage treatment infrastructure to prevent untreated wastewater from being dumped into rivers, lakes, and oceans. This has improved the level of sewage treatment substantially in recent decades.

There has also been progress in the developing world, due to government investments in infrastructure and funding from international aid. Non-government organizations have also played a major role. For example, Water.org provides money to microcredit agencies that have extended more than $70 million in loans for water and sanitation to people in poor countries. Their success is incredible: almost two million people have benefitted; 91 percent of these loans have gone to women; and the repayment rate is 99 percent.

However, more than two billion people still lack access to adequate sanitation facilities. More than a billion people, the majority of whom live in India, must still defecate in fields, forests, bushes, bodies of water, or other open spaces. Some still use the flying toilet, a euphemism for defecating in a plastic bag and then throwing it away. These practices spread germs, pollute water, and can contaminate agricultural products. While there has been tremendous progress in securing access to safe drinking water, the issue of sanitation has lagged behind, and the goal of cutting the proportion of people without access to adequate sanitation in half by 2015 was not met.

PROGRESS IN PROVIDING safe drinking water and sanitation is matched by giant strides in reducing hunger. Back in 1798, Thomas Malthus predicted that food production would be unable to keep up with a growing human population, resulting in widespread starvation. In 1968, Paul Ehrlich made a similar forecast in his book *The Population Bomb*, predicting that mass starvation would kill hundreds of millions of people in the 1970s because of overpopulation. Malthus and Ehrlich have been proven wrong. Globally, malnutrition and undernutrition are declining despite human population growth.

The rapid increase in food production comes with a major caveat. Industrial agriculture has had devastating environmental impacts, including the destruction of native grasslands, forests, and wetlands (all contributing to loss of wildlife habitat); loss of seed biodiversity; water pollution caused by pesticides, fertilizers, and manure; and emissions of climate-changing gases. We may be able to feed seven billion people, but not on the super-sized, meat-heavy North American diet.

Although the globalized food industry continues to pose major environmental challenges, there are some encouraging signs, such as rising popularity of organic agriculture, local foods, and fair trade products. Public pressure has led to new regulations requiring more accurate, informative, and understandable food labelling as well as the elimination of unhealthy ingredients from some products. Many people are adopting healthier diets, generating benefits for both themselves and the environment.

Many people assume that government regulations and food inspectors ensure that what is sold in grocery stores and restaurants is safe to eat. Far too often, this is not the case. Although food safety standards have improved over the years, there are still many gaps and glitches. In 2014, CBC's *Marketplace* television program asked me to appear on their show because they had tested a number of tea products for pesticide residues and wanted an environmental lawyer's response to the results. Tea is one of the world's most popular beverages, with Canadians sipping an estimated ten billion cups per year. The CBC tested Canada's most popular brands and was astonished to find pesticides above the legal limit on half the samples. One tea imported from China had residues from 22 different pesticides. These pesticides are a threat to farmers, the environment, and consumers. Yet the government responded by shrugging its shoulders, refusing to take any enforcement action despite the violations of Canadian law.

In light of many similar stories, it is no surprise that organically produced food is the fastest growing sector of the global food industry. Even during the recent recession, sales of organic food jumped by 10 to 15 percent per year. Organic food, grown without toxic chemicals and fertilizers, antibiotics, growth hormones, or genetically modified organisms, reduces foodborne health risks and is better for maintaining and enhancing soil and ecosystems. A recent scientific analysis of 343 peer-reviewed articles concluded that organic foods offer higher concentrations of antioxidants (which reduce risks of heart disease, neurodegenerative diseases, and cancer) and, not surprisingly, lower levels of heavy metals and pesticide residues. The value of the Canadian organic market has tripled since 2006, with more than half of all Canadians purchasing organic foods on a weekly basis. In the U.S., organic food sales are increasing 10 to 12 percent annually, while eight in ten American parents report that they try to purchase organic products despite higher prices.

Jim Cochran is California's first commercial organic strawberry grower and owner of Swanton Berry Farm, near Santa Cruz. Cochran, unlike the majority of commercial berry farmers, has grown organic berries for 30 years. When he was a boy, Cochran's grandmother served him his favourite treat: a big bowl of little strawberries. The fresh-picked berries were sweet and exploded with flavour. "They made a big impression on a five-year-old," said Cochran. As a young man, Cochran worked on a farm that sprayed strawberries with acutely toxic pesticides including chloropicrin and methyl bromide. Chloropicrin was used as a poison gas during the World War I. Methyl bromide, like CFCs, contributes to the depletion of the ozone layer and was supposed to be totally phased out years ago, but industrial agriculture interests in Canada and the U.S. have lobbied for exceptions that continue to this day. Other farmers told Cochran that strawberries could not be grown commercially without pesticides because of their susceptibility to mould, soil diseases, and other problems. One day, he was working in a recently sprayed strawberry field and breathing pesticides made him ill. That was his turning point, and he began to pursue the organic path. His farm was certified organic in 1987 and has won numerous green agriculture awards.

Another positive trend has been skyrocketing support for locally produced food. The *Oxford Dictionary* word of the year in 2007 was locavore, meaning a person who eats locally grown and produced food whenever possible. Farmers' markets in Canada and the U.S. more than tripled in number over the past 20 years and are wildly popular. Farmers' markets promote food education and create bonds between producers and consumers. At the Pender Island farmers' market, held on Saturday mornings, keen shoppers begin lining up 45 minutes before the opening bell in order to have first dibs on the fresh produce. Food sold at these markets has less packaging and travels a shorter distance than most of the food in a grocery store, which generally reduces its carbon footprint. Buying local food pumps more money into the local economy, supporting jobs in the community. In many cities, community gardens are enjoying explosive growth in popularity. Vancouver has added more than 1,200 community garden plots in recent years, bringing the total close to 5,000. Community orchards (up from three in 2010 to 33 today in Vancouver) and edible landscaping are also taking off. Classes are being offered in gardening, composting, cooking, and food processing. Cities are also working to eliminate food deserts—areas where there is a shortage of healthy foods available for purchase. Detroit and L.A. are examples of cities where urban gardening is flourishing in low-income neighbourhoods.

Eagle Creek Farms in central Alberta, where my late Uncle Bill and Auntie Anne used to raise cattle and grow grain on hundreds of hectares of rolling prairie, exemplifies the food revolution that is underway. Their son Stan, my cousin, ran the family farm for decades but now has handed most of the reins to his youngest son, John, who has wanted to farm since he was little. (I remember him asking for real tools like shovels and rakes for Christmas as a wee boy.) The cattle are gone and most of the fields are leased to other farmers. Eagle Creek Farms now offers a community-supported agriculture program. Community-supported agriculture creates a relationship between farmers and consumers, in which the latter pre-pay for a season of fresh, local, and tasty food, sharing the risk of crop failures, bad weather, and other catastrophes. Farmers deliver a weekly box of mixed produce for as many weeks as possible. There are now thousands of CSA programs in North America and Europe.

John's farm has fields of u-pick flowers and vegetables, corn and sunflower mazes, a petting zoo, and the CSA service providing over 90 different varieties of produce—from strawberries to kale—to almost 400 families. He also sells produce at a farmers' market in Calgary. All of this food is grown without synthetic chemicals, pesticides, or artificial fertilizers.

John is also building some innovative passive solar greenhouses, using Freecycled materials and thick walls to retain the sun's heat. These off-the-grid greenhouses also have rainwater collection systems. He hopes to be able to grow some hardy crops like kale for 10 to 11 months each year despite Alberta's cold winters. An inspiration for John's efforts is the home of energy guru Amory Lovins, where bananas grow indoors despite the building's location at 2,400 metres above sea level in Snowmass, Colorado.

COMPLEMENTING THE RISING COMMITMENT to organic and small-scale local agriculture are fair trade foods, certified to ensure that farmers receive fair compensation for their efforts. Fair trade proponents define fair trade as a partnership based on dialogue, transparency, and respect, that seeks greater equity in international trade. Fairtrade International has certified more than 27,000 different products. Two of my favourite fair trade products are Alter Eco quinoa and chocolate. Alter Eco partnered with a Bolivian farmers' cooperative called ANAPQUI, whose membership includes 1,100 small-scale farmers. Alter Eco has contributed to bringing running water and sanitation to 15 Bolivian villages as well as the region's first high school and a health care clinic. The company estimates that its partners receive an additional 30 percent in revenue compared to other Bolivian quinoa farmers. Cadbury now makes its famous Dairy Milk bar with fair trade chocolate, paying a premium to small-scale farmers in Ghana and other African countries.

MANY OF THE POSITIVE CHANGES to the food system are being driven by public demand and in some cases the unprecedented power of social media. A number of recent campaigns have targeted toxic substances that threaten both human health and the environment. Bloggers forced fast food chain Subway to stop using azodicarbonamide, a plastic-based additive, in the breads that they serve. Missouri teenager Sarah Kavanagh led the effort to convince both Coca-Cola and Pepsi to reformulate their flagship sports drinks to eliminate the use of brominated vegetable oils, an ingredient closely related to flame retardants used in furniture and electronics. Vani Hari drummed up tens of thousands of signatures on a Change.org petition demanding that Kraft remove artificial yellow food dyes from its macaroni and cheese. The dyes, linked to adverse health effects, were used only in the North American version of the product. Houston healthy food advocate Bettina Siegel has worked to have "pink slime" removed from American school lunches. Pink slime, what the meat industry calls "lean finely textured beef," is a filler made by putting meat by-products in a centrifuge and then treating it with ammonia to kill harmful pathogens such as *E. coli* and *Salmonella*. Siegel is also fighting to prevent American-raised chicken from being sent to China for processing and then returned to the U.S. for sale, a practice that would require burning huge amounts of fuel for transport. Environmentalists have convinced municipal governments in New York, Los Angeles, San Francisco, Seattle, Portland, and other cities to ban foam fast-food packaging, passing bylaws that require recyclable or compostable substitutes.

Consumers and health advocates have demanded more accurate and informative food labelling to address genetically modified organisms; excessive levels of sugar, fat, and salt; and the use of artificial colours, flavours, and sweeteners. More than 60 countries have implemented mandatory GMO labelling, including the entire European Union. In many of these countries, foods containing GMOs are no longer sold, as consumers avoid buying them when informed of their presence. Canada changed rules that previously permitted apple juice made from Chinese apples to be labelled "Product of Canada" if the processing was done in Canada. Europe has banned the use of a number of food dyes linked to hyperactivity and attention deficit disorder in children and requires warnings for parents on foods containing other potentially harmful substances. Knowledge is definitely power in the context of information about what we eat.

HEALTHIER DIETS ARE A KEY ASPECT of reducing the environmental pressures created by industrialized agriculture. I grew up on a typical North American diet, eating a lot of meat. Thanks to my generous Uncle Bill, we always had a freezer full of beef, so we ate hamburgers, tacos, sloppy joes, chili, spaghetti and meat sauce, beef stew, and meatloaf. We even ate tongue sandwiches occasionally. As a skinny teenager, confident that my appetite was insatiable, I entered a hamburger-eating contest at a Calgary restaurant. First prize was a trip to Hawaii, a place I dreamed of visiting. I was the youngest and scrawniest of the 25 contestants. We were given 30 minutes to eat as many burgers (bun included, condiments optional) as possible. Copying the strategy of my burly competitors, I pounded the burgers with my fist to flatten them. At the halfway point, I was in the lead, having wolfed down eight and a half burgers. Visions of beaches and surfing danced in my head. On my twelfth burger, I had difficulty swallowing. The unthinkable had happened. I felt the food go partway down my throat and stop. I sipped some water but it made no difference: my stomach was completely full and food was backing up into my esophagus. As the clock wound down, I sat there stunned as two gargantuan men passed my total, relegating me to third place. It was two days before I could eat a normal meal, and burgers never again held the same attraction. North Americans may not go to the same extreme, but most people need to reduce meat consumption for both health and environmental reasons.

I became a vegetarian before becoming an environmental lawyer. The spark came when I started doing triathlons (swim-bike-run endurance events) and learned that the greatest triathlete in the world was a vegetarian. Dave Scott had won the world Ironman championship in Hawaii a record six times, putting to rest any concerns I had about the health consequences of giving up meat. In the ensuing 25 years, my weight has not changed, my health is still great, and I continue to compete in long-distance triathlons and ultramarathons. Margot and Meredith are vegetarians as well, though we will occasionally eat a local fish taco, wild salmon, or Pacific halibut.

Research indicates that a vegetarian diet results in longer life expectancy and less risk of heart disease, obesity, and cancer. It can also reduce the environmental impact of one's diet by as much as 90 percent. Current industrial agricultural practices for raising cattle, pigs, and chicken require up to ten times as much water and energy and produce up to ten times the greenhouse gas emissions of plant protein. There are now more than a billion people who rarely or never eat meat, and the number continues to grow. But a healthy, sustainable diet can still include animal protein. You just need to be selective about the meat you buy, seeking out advice from local farmers, ranchers, or butchers who can provide meat raised in a more sustainable, and humane, way. As acclaimed food writer

Michael Pollan concludes in *The Omnivore's Dilemma*, the best diet advice for the health of people and the planet can be summed up in seven words: "Eat food. Not too much. Mostly plants."

IT IS HEARTENING TO SEE the remarkable progress being made in regard to environmentally sustainable provision of water, sanitation, and food. More people than ever before are enjoying access to these prerequisites of a good life. To be sure, renewed efforts are required to fulfill the needs of those who still lack access to safe drinking water, indoor toilets, and adequate nutrition. Further strides can and must be made; however, it's essential that further progress for people not come at additional cost to the planet.

*"I am fundamentally an optimist.
Whether that comes from nature or nurture, I cannot say.
Part of being optimistic is keeping one's head pointed toward the sun,
one's feet moving forward. There were many dark moments when
my faith in humanity was sorely tested, but I would not and could not
give myself up to despair. That way lays defeat and death."*
—Nelson Mandela

Chapter 7
Global Detox

I HAVEN'T HAD MY BLOOD TESTED to determine the presence of toxic substances in my body, but I know that they are there, despite my best attempt to live a clean, green, healthy lifestyle. Our family lives in a community with minimal air pollution. Our water comes from a well drilled deep into an underground aquifer and is both safe and delicious. We eat a primarily local, organic, and vegetarian diet. We purchase non-toxic consumer products, from cleaners to paints. And yet I know that my body, Margot's body, and even Meredith's nine-year-old body contain a witches' brew of pesticides, stain repellants, flame retardants, plastic softeners, heavy metals, and other industrial chemicals. Extensive studies conducted on thousands of people in North America and Europe have found these substances in everyone. This bad news has received extensive media coverage and as a result is fairly well known. The good news, which far fewer people are aware of, is that the levels of many of these toxins within us are declining. There is uncertainty about whether, and to what extent, the low levels of chemicals found in people's bodies cause adverse health effects. However, numerous examples including lead, benzene, and air pollution show that we often underestimate the risks associated with toxins. Acceptable levels of exposure to these substances have been lowered time and time again by scientists and government agencies.

For their book *Slow Death by Rubber Duck*, two friends of mine, Rick Smith and Bruce Lourie, deliberately exposed themselves to a cocktail of seven toxic substances encountered through food and consumer products in the course of ordinary daily life including stain repellents, bisphenol A, phthalates, triclosan, flame retardants, pesticides, and mercury. They managed to boost their body burdens of these toxic chemicals by doing things like locking themselves in an apartment where the furniture was recently treated with stain repellents, eating meals microwaved in plastic containers, and dousing themselves with heavily scented soaps, shampoos, and other toiletries. For their sake, I'm glad they didn't pull their stunts a generation earlier because the toxic substances encountered in daily life back then, such as lead and earlier generations of pesticides, were much more dangerous.

For decades, scientists synthesized new chemicals that promised a miraculous array of uses. In the 1950s, ads touted "better living through chemicals" and plastics were plugged as the fifth food group. Corporations sold these chemicals in vast quantities. Decades later, however, scientists uncovered unexpected and potentially devastating health and environmental impacts caused by these toxic substances. DDT went from wonder chemical to global pariah when it was identified as the cause behind the disappearance of bald eagles and other birds at the top of the food chain.

CFCs were a modern marvel until it was revealed that they were chomping Pac-Man-like through the Earth's protective ozone layer. DDT and CFCs both made *Time* magazine's list of the 50 worst inventions of all time, along with relatively benign creations such as the vibrating ab belt, Tamagotchis, and hair in a can.

In the 1980s, it was revealed that pesticides banned in the U.S. for health and environmental reasons were being exported to other countries, where these toxic substances were then used to produce agricultural products exported back to the U.S. Farmworkers in developing countries were put at risk by exposure to these pesticides, and residues of these banned pesticides were detected on products being sold to American consumers. A disturbing example was DBCP, a pesticide banned in the U.S. after men working in a factory manufacturing the chemical became sterile. However, the food giant Dole continued to use DBCP on banana plantations in Latin America for years after it was banned in the U.S., injuring thousands of unsuspecting agricultural workers.

These repugnant practices, and catastrophes like the disaster in Bhopal, India, where a chemical release from a pesticide factory killed thousands of people, led to the negotiation of the Rotterdam Convention in 1998. This treaty limits trade in more than 50 toxic industrial chemicals, requiring importing countries to provide their prior informed consent before the listed toxic substances can be shipped.

In 2001, the Stockholm Convention on Persistent Organic Pollutants banned 12 of the most toxic substances: DDT, aldrin, dieldrin, endrin, heptachlor, hexachlorobenzene, chlordane, mirex, toxaphene, dioxins, furans, and PCBs. These chemicals remain in the environment for years, accumulating in ecosystems and wildlife on land and in the water. In recent years, more substances—including pesticides, brominated flame retardants, and other industrial chemicals—have been added to the Stockholm Convention's coverage. Proposals to add another five chemicals to the convention are currently under review.

During the 1980s, there were a number of high-profile incidents where hazardous waste from rich countries was dumped in Eastern Europe or developing nations. For example, in 1986, a ship carrying incinerator ash from Philadelphia was found illegally dumping half its cargo on a beach in Haiti. The ship was forced away from Haiti and became an international pariah, changing its name and sailing around the world in search of a willing recipient for its waste. Having failed to find a taker, the ship eventually dumped the waste on the high seas in the Indian and Atlantic Oceans. Similarly, in 1988, a flotilla of ships from Italy carrying 8,000 barrels of hazardous waste dumped its cargo in the Nigerian community of Koko, paying a farmer $100 monthly for "storage."

These incidents were spurred by increasingly stringent rules about the management of toxic waste in wealthy industrialized countries and the higher costs of proper disposal that ensued. In response to these dumping scandals, the nations of the world negotiated the Basel Convention in 1989. This international treaty places some restrictions on trade in hazardous wastes by requiring notice, consent, and tracking the movement of these wastes. A subsequent amendment, not yet in force, would prohibit all countries belonging to the Organization for Economic Cooperation and Development and the European Union from exporting hazardous waste to developing countries. This Basel Ban Amendment is opposed by a handful of countries including

Canada, the U.S., and Australia, for cynical reasons, such as complaining that it doesn't apply to hazardous waste shipments from poor countries to wealthy countries. Unhappy with the delay, the European Union passed its own law completely banning hazardous waste exports from its 28 members to developing countries.

SINCE THE 1950S, a group of organotin compounds (mixtures of tin and hydrocarbons), including tributyltin (TBT), were added to antifouling paint used on boats to prevent organisms from attaching to the hull. TBT was thought to be less toxic to marine life than the biocides in use at that time, including DDT and arsenic. Unfortunately, it turned out to be highly hazardous to the environment, wildlife, and possibly humans. It persists in the ocean for up to 30 years, accumulating in the food chain, and is one of hundreds of chemicals that interfere with the endocrine system that regulates hormones in animals, including people. Scientists describe TBT as "the most toxic substance ever deliberately introduced into the marine environment." The use of TBT in antifouling paint was prohibited in some countries as early as the 1980s, while a global ban came into effect in 2008. In 2014, it was reported that penises on female snails are disappearing, an improvement linked to the phaseout of TBT. That is what I call an uncomfortable piece of good news. It makes me squirm. Appallingly, exposure to TBT had been wreaking havoc on snails' endocrine system, causing the penises to develop. In another surprising scientific discovery, researchers investigating the reproductive effects of TBT exposure on frogs found that the animals became grossly obese, leading the scientists to conclude that exposure to hormone-disrupting chemicals may play a role in the growing global obesity epidemic.

Manmade chemicals are not the only potential sources of serious harm to human and ecosystem health. Mined metals and minerals, such as mercury, lead, and asbestos are also potentially toxic. When I was a little boy, a friend of mine had a vial of what we thought was the coolest substance ever—far more fascinating than Play-Doh or silly putty. It was shimmering and silver and would roll around on your hand like a ball bearing, except globules could separate and then rejoin. Liquid at room temperature, yet a metal. It was mercury, a heavy metal that, unbeknownst to us at the time, is highly toxic and can impair normal neurological development.

Despite mercury's toxicity, governments and industries dragged their heels in addressing this danger. In the 19th century, it was understood that hat makers suffered chronic poisoning caused by breathing mercury vapors in the felting process, leading to the expression "mad as a hatter." Given that, it's pretty shocking that it took until 2013 to conclude negotiations on the Minamata Convention, an international treaty aimed at reducing the use and release of mercury. The treaty is named after the Japanese city where a factory dumped mercury into a bay, poisoning seafood that people then consumed. Hundreds of people died and thousands more suffered debilitating damage to their brains, kidneys, and lungs. The convention bans new mercury mines, phases out existing mines, controls emissions to the air, and regulates the use of mercury in small-scale gold mining. It builds on the substantial progress that has been made by some countries over the past two decades in reducing the use and release of mercury. Mercury emissions have fallen by 90 percent in Canada since 1990, mainly due to reductions in the smelting and refining industry. Emissions from burning coal to generate electricity, which releases mercury as a by-product, are also down by 60 percent. In the U.S., mercury emissions from incinerators are down by more than 90 percent, while new rules for coal-fired power plants should achieve similar reductions in the electricity sector by the end of 2015.

UNTIL THE EARLY 1970S, signs at gas stations had two prices, one for leaded gasoline and another for unleaded. The latter was sold for newer vehicles equipped with improved technology. Yet adding lead to gasoline was one of the dumbest ideas of the 20th century. Cars using leaded gasoline spewed tiny particles of lead from their tailpipes, contaminating the air and eventually, when the particles settled, the soil. People, including children, inhaled that air and played in the dirt, inadvertently exposing themselves to a damaging neurotoxin. Lead exposure also was caused by diverse other sources, including paint, plumbing, ammunition, and costume jewellery. Incredibly, lead poisoning had been recognized since the time of ancient Rome.

Exposure to lead, easily measured through blood samples, has irreversible impacts on development and intelligence. The higher the level of childhood lead exposure, the greater the level of IQ loss lasting a lifetime, with devastating social and economic costs. A handful of IQ points can make the difference between low but functional intelligence and a lifetime of dependence, or the difference between very smart and the level of genius that could solve some of the world's most vexing problems.

According to a 2013 report from Health Canada, early childhood exposure to lead in Canada causes a decrease in future earnings due to reduced intellectual development that costs between $1.5 billion and $9.4 billion annually. Childhood lead poisoning in the U.S. still inflicts costs of over $50 billion per year, a calculation based on lower IQs and reduced lifetime earnings. Every dollar spent to reduce lead hazards in housing is expected to produce between $17 and $221 in benefits by reducing expenditures on special education and screening and treatment for lead exposure and ADHD; increasing income and tax revenue; and reducing crime.

As scientific evidence about the harmfulness of lead grew stronger, authorities repeatedly lowered the threshold for what is considered a safe level of exposure—from 60 micrograms per decilitre of blood in the 1960s to 40 in 1971, 30 in 1975, 25 in 1985, 10 in 1991, and 5 in 2012. Today, leading medical experts believe there is no safe level of exposure.

The elimination of leaded gasoline caused ambient levels of lead in the air to fall by more than 90 percent. Thanks to the elimination of leaded gasoline and regulations limiting lead in paints and other products, average lead concentrations in the blood of American and Canadian children have declined by more than 90 percent since 1980. There are still lead pollution hotspots near smelters. Older homes may pose a risk to young children where paint containing lead is peeling or pipes leach lead into drinking water. There is still a handful of countries that have not phased out leaded gasoline, and one irresponsible corporation, the U.K.'s Innospec, that still sells lead to these countries.

Interestingly, in country after country, as blood lead levels rose, so did rates of crime. Leaded gasoline would be phased out, and a few years after that, rates of crime would plunge. Since countries experienced the rise and fall of lead exposures and changes in criminal activity at different times, the consistency of this association is striking. Researcher Rick Nevin studied blood lead levels and crime rates in Canada, the U.S., the U.K., France, Germany, Australia, Finland, Italy, and New Zealand and discovered a strong correlation in each country. Scientists searching for explanations point towards new evidence about lead exposure and its effects on parts of the brain responsible for impulse control.

Some European countries banned lead-based paints in the early 20th century, while Canada and the U.S. took many decades to follow suit because of intense and misleading lobbying from paint manufacturers. Lead-based paints and leaded gasoline are still sold in some developing countries, often by corporations that make lead-free products for other markets. American paint corporations are now being forced to pay billions of dollars in damages for their negligence. In a California lawsuit brought by ten cities, a judge ordered three major paint companies to pay $1.1 billion for the removal of hazardous lead paint from millions of homes.

ASBESTOS WAS ONCE REGARDED as a miracle mineral because it is strong, lightweight, and heat resistant. Over the course of the 20th century, asbestos was widely used in construction materials such as roof tiles, ceiling tiles, insulation, and pipes, in brake pads for vehicles, and in an array of consumer products from oven mitts to toys. Breathing asbestos fibres creates a serious risk of deadly diseases including asbestosis, lung cancer, and mesothelioma. These illnesses do not appear until decades after exposure and are incurable, resulting in a terrible quality of life and painful death. Asbestos is still the biggest workplace killer in Canada, the U.S., the U.K., and New Zealand.

Remarkably, Canada and the U.S. continue to allow the use of asbestos in many different applications. The Canadian government, despite the recent demise of the domestic asbestos mining industry, continues to insist that it can be used safely. A 2014 investigation by the *Globe and Mail* revealed that asbestos pipe was being imported into Canada and used in new construction ranging from condominiums to hospitals. At the same time, highly specialized hazardous waste businesses are being paid handsomely to remove asbestos pipe from existing buildings. In fact, many millions of dollars have been spent removing asbestos materials from Canada's Parliament buildings. Tom Kelly, president of the asbestos removal company Inscan Kaefer Inc., said it is incongruous that his business is removing exactly the same materials that are still being installed. Canada was once among the world's leading exporters of asbestos, and an avid cheerleader as well. Now that the last asbestos mine in Canada has closed, the federal government finally ended its decades-long, multimillion-dollar support for the industry's public relations organization. In the U.S., a law banning asbestos was overturned by the courts in a lawsuit filed by the industry. On the other hand, American asbestos companies have paid billions of dollars in compensation to victims and their families, with expert estimates pegging the eventual total in excess of $200 billion.

More than 50 countries have banned all uses of asbestos, including Australia, Japan, and all 28 countries in the European Union. The World Health Organization has stated unequivocally that all uses of all types of asbestos should be eliminated. Asbestos production has fallen from a peak of five million tonnes annually in 1980 to two million tonnes today, but this means that thousands of people, primarily in developing countries, are still being exposed to a substance that will cause disease and death decades down the line.

DEBATES ABOUT CHEMICAL SAFETY are often David and Goliath battles featuring a few courageous activists, scientists, and doctors on one side, and giant corporations, industry associations, and their government cronies on the other. The amazing thing is how often the Davids emerge victorious. Industries whose products come under attack routinely follow a game plan based on tobacco industry tactics. First they deny the existence of any problems. They pay charlatan scientists to lie and say their products or emissions are safe. They finance scientific journals with official-sounding titles to publish bogus articles based on junk science. They wield their wealth in efforts to buy the support or acquiescence of politicians and bureaucrats.

There was substantial scientific evidence in the 1920s and 1930s that lead, mercury, and asbestos posed serious threats to human health, yet these toxic substances continued to be widely used for many more decades. Industry-funded research on the deaths of asbestos miners and the neurological impacts of lead exposure was suppressed. Health problems were covered up. In the words of British journalist Nic Fleming, the story is "one of scientific deception and betrayal, greed, political collusion, the power of propaganda, and above all, the willingness of some executives to knowingly subject hundreds of thousands of people to severe illness and even death in the pursuit of profit." In their book, *Defending the Indefensible*, professors Geoffrey Tweedale and Jock McCulloch conclude, "Each time the evidence that industry lobbyists were wrong became overwhelming, they just changed tack."

Polybrominated diphenyl ethers represent a recent and prototypical example of the ongoing debate about the use of chemicals in modern society. In the 1950s, highly flammable synthetic fabrics began replacing less flammable natural fibres. In response, PBDEs were synthesized for use as flame retardants in products ranging from children's pyjamas to furniture. They were believed to be safe: essential for fire safety and non-toxic. Natural processes were supposed to break these compounds down into even more harmless substances. They were not expected to enter the air, water, or soil, accumulate in living organisms, or build up in the food chain. These industry claims turned out to be wrong, but refuting them took decades. Under American chemicals law, substances already in commerce prior to 1976 were grandfathered, meaning no safety testing was required.

Swedish scientists first found traces of PBDEs in women's breast milk in the early 1990s. These chemicals were not supposed to appear in humans. Alarmed, researchers in Sweden and other countries began regular monitoring. Levels of PBDEs in women's breast milk in other wealthy countries skyrocketed. These supposedly benign chemicals were migrating from consumer products into air, water, and food, entering our bodies, and accumulating there. Not good. Although the human health impacts of exposure to PBDEs are not well understood, tests on animals indicate impaired brain development, negative impacts on the endocrine and reproductive systems, and possibly cancer. New research found that higher levels of PBDEs in babies' cord blood are associated with developmental delays in children aged one to six.

Americans had the highest recorded PBDE concentrations in the world. Canadians ranked second. In Vancouver, PBDE levels measured in breast milk increased by 1,500 percent between 1992 and 2002. North American children have even higher PBDE concentrations than their parents. Foods in Canada and the U.S.—including ground beef, cheese, and butter—were contaminated with PBDEs at levels up to 1,000 times higher than similar food products in Europe. Scientists found burgeoning levels of PBDEs in a wide range of animals at the top of various food chains—from killer whales to grizzly bears. Canada's leading marine mammal toxicologist, Dr. Peter Ross, described orcas as "fireproof whales" because the concentrations of PBDEs in their bodies were so elevated.

Sweden's government, an internationally renowned leader in chemical regulation, acted fast to ban PBDEs and began calling for other nations to do the same. Levels of PBDEs in the breast milk of Swedish mothers rapidly declined after the government began taking action in 2001. Studies also show that body burdens of dioxins in Swedes have dropped by more than 90 percent over the past 30 years and PCB levels are also falling. The Swedish experience illustrates the health benefits of strong environmental regulation.

In the U.S., Republicans have delayed effective action on all chemical regulation at the national level for many years. In response, environmentalists and health activists sought restrictions on PBDEs and other flame retardants at the state level. A dozen states, led by Maine and Vermont, enacted laws restricting the use of some PBDEs. California banned the use of two main PBDEs in 2003. By 2010, concentrations of PBDEs in pregnant women tested in San Francisco dropped by two-thirds.

California became a key battleground because the state had a 1975 law in place that required furniture manufacturers to treat their products with flame retardants. New research from independent scientists and the U.S. Consumer Product Safety Commission concluded that PBDEs and other flame retardants provide no meaningful protection from fire. Advocates pushing for the elimination of the mandatory flame retardant rule were making progress. Firefighters joined the campaign because they were concerned about the health impacts of breathing in the fumes of these toxic substances. The chemical industry was on its heels. But unexpected testimony from a highly respected medical doctor turned things around for the pro–flame retardant campaigns in California, Alaska, and Washington.

In front of lawmakers, Dr. David Heimbach testified that he recently had treated a six-week-old baby with half of her body covered in severe burns caused by a candle fire and the fact she was sleeping on a pillow that did not contain flame retardants. Heimbach said, "She ultimately died after about three weeks of pain and misery in the hospital." Members of the audience literally gasped. It was emotional, compelling testimony and was enough to change the minds of some California lawmakers who'd been leaning towards a ban. Heimbach told lawmakers in Alaska and Washington similar stories, although the details varied from state to state. Thanks to his testimony, the bills seeking to limit the use of flame retardants were defeated in all three states.

But there was an intriguing twist to the story that took several years to surface. Dr. Heimbach's tear-generating stories were fabricated. An exemplary piece of investigative journalism by reporters with the *Chicago Tribune* revealed that Heimbach was an industry stooge, paid $240,000 to testify by a bogus organization called Citizens for Fire Safety. This group claimed on its website to be "a coalition of fire professionals, educators, burn centres, doctors, fire departments, and industry leaders, united to ensure that our country is protected by the highest standards of fire safety." In fact,

Citizens for Fire Safety was set up by three of the largest manufacturers of flame retardants and had no other members.

After Heimbach's fraud was unmasked, the California law was finally amended in 2013, meaning furniture in that state can be made and sold without being doused with flame retardants. The California precedent is expected to have wide-ranging implications in other jurisdictions, both in North America and beyond. Washington State medical authorities charged Heimbach with fabricating testimony and falsely representing himself as an unbiased burn expert. He lost his licence to practice medicine. The group Citizens for Fire Safety no longer exists. The *Chicago Tribune* concluded, "A decades-long campaign of deception has loaded furniture and electronics in America with pounds of toxic chemicals linked to cancer, neurological deficits, developmental problems, and impaired fertility." The flame retardant story offers vital lessons for future debates about the safety of chemicals.

WITH CONSISTENT ENCOURAGEMENT from Scandinavian countries, the European Union now leads the world in taking a proactive and precautionary approach to chemicals management. In addition to being a leader on substances ranging from asbestos to PBDEs, there are dozens of pesticide ingredients, used in more than 1,000 commercial pesticide products available in Canada and the U.S. that cannot be sold in the EU because of health and environmental concerns. For example, in 2004, the EU banned atrazine, which is known to wreak havoc on the reproductive systems of frogs exposed to small concentrations. Yet atrazine is among the most heavily applied pesticides in North America, used widely on corn and soybean crops. In 2012, atrazine's manufacturer settled a lawsuit with a number of municipal governments in the U.S., compensating them to the tune of over $100 million for contaminating their water supplies.

In recent years, new medical research has indicated that artificial food colourings such as Red Dye No. 40, Yellow Dye No. 5, and Yellow Dye No. 6 cause increased hyperactivity in children. The European Union requires warning labels on foods containing these artificial colours to state that they "may have an adverse effect on activity and attention in children." This regulatory step has greatly decreased their use in Europe. Yet the use of these dyes as food additives continues to be permitted in Canada and the U.S. without any warnings. Similarly, the EU took a much more proactive stance in response to scientific studies that found elevated levels of arsenic in a range of food products made with rice, increasing the risk of cancer. The EU encouraged people who regularly consume rice to alter their diet to include a diverse mix of grains and warned that certain populations were particularly vulnerable. In contrast, Canada and the U.S. determined that the elevated levels of arsenic posed no threat to the health of any consumers and took no action.

Occasionally there are flashes of leadership from North America. Canada received international accolades for being the first country to ban the use of bisphenol A (BPA) in baby bottles. BPA is used to make plastics for a wide range of consumer products. Canada also implemented pioneering regulations for a group of toxic substances called perfluorochemicals. PFCs are used in many industrial processes as well as in consumer products including fire-fighting foams, household cleaners, cosmetics, electronics, food packaging, non-stick coatings on pots and pans, and stain repellents on furniture and clothing. They are linked to cancer, birth defects, and damage to the liver, pancreas, immune system, and reproductive system. The Environmental Protection Agency leaned heavily on major manufacturers of certain flame retardants and stain repellants and persuaded them to voluntarily phase out production, both within the U.S. and globally. Massachusetts passed a strong law called the Toxic Use Reduction Act in 1989. By 2010, releases of 74 known or suspected cancer-causing chemicals from facilities in the state had fallen by an average of 93 percent compared to 1989 levels. For example, perchloroethylene and trichloroethylene, carcinogenic chemicals used in dry-cleaning and metal degreasing processes, decreased by 85 percent and 92 percent respectively.

THERE IS TANGIBLE PROOF that the international treaties and national laws described in this chapter are working to protect the health of humans and ecosystems. Levels of many of these deadly chemicals found in people, other species, and the environment are beginning to decline. The U.S. has been comprehensively testing the chemical body burden of citizens since the late 1990s. A report released in 2014, based on the most up-to-date data available, showed declines in many toxic chemicals found in Americans' blood, including bisphenol A, lead, mercury, cadmium, xylene, the gasoline additive MTBE, 2,4-D, chlorpyrifos, some perfluorochemicals (e.g., PFOA and PFOS), and some phthalates (e.g., BBP, DEP, DEHP). For many other substances, levels were below the limit of detection, meaning sophisticated tests found no traces of these chemicals in people's bodies. Similar declines are probably occurring in Canada and Europe. The decline in body burdens means that people's exposure is declining, indicating lower levels of these toxic substances in the air we breathe, the water we drink, the food we eat, and the consumer products that we purchase.

The declining levels of toxic substances in the environment also benefit plants, wildlife, and ecosystems. The cleanup and restoration of False Creek, a narrow saltwater inlet in the heart of Vancouver, enabled the return of herring to spawn. The return of herring was followed by the appearance of a gray whale in downtown waters. On the other side of the continent, improved water quality in the Hudson River is credited with growing numbers of humpback whales and great white sharks frequenting the harbour adjacent to New York City. A new study of chemical concentrations in marine mammals around the world shows many contaminants that have been subject to regulation are declining, including organochlorine pesticides such as DDT, PBDEs, and tributyltin (TBT). These largely unheralded decreases in levels of toxic substances found in humans and wildlife are a highly encouraging sign of a society that is finally treating the environment with greater respect. We now seem to understand the old adage that when we damage the environment, we are also harming ourselves.

Part III
THE BUILT ENVIRONMENT

*"The average pencil is seven inches long, with just a half-inch eraser—
in case you thought optimism was dead."*
—Robert Brault

Chapter 8
The Greenest City Decathlon

FOR THE FIRST TIME IN HISTORY, more than half of the people in the world now live in cities. By 2050, 70 percent of humanity, or 6.4 billion people, will live in urban areas. Not surprisingly, cities already account for almost three-quarters of total resource use and 80 percent of global greenhouse gas emissions. In light of these facts, redesigning cities to be sustainable is among our most urgent challenges. I'd been reading these and similar statistics for years, but it was a list of the world's most populous cities by 2100 that really drove the point through my admittedly thick skull. Topping the list by the end of the century are Lagos, Kinshasa, and Dar es Salaam, while others in the top 25 include Niamey, Lilongwe, and N'Djamena. If you can place these cities on a map, then your geography knowledge is far better than mine. Noticeably absent are Tokyo, Mexico City, and Beijing, today's most populous urban areas.

Fortunately, there is a global race among many metropolises, each striving to be recognized as "the greenest city in the world."

This claim has been staked by a diverse array of cities including Stockholm, London, Freiburg, Copenhagen, Oslo, Amsterdam, Berlin, Portland, New York, Chicago, Chandigarh (India), Curitiba (Brazil), and even Calgary (an unlikely contender as the oil and gas capital of Canada and a legendary example of urban sprawl). Brand new "eco-cities" are being built in places as diverse as Portugal, China, Qatar, and the United Arab Emirates. In 2009, the mayor of Vancouver, Gregor Robertson, asked me to join him in co-chairing the Greenest City Action Team, with a mandate to propel his city to the forefront of the world's most environmentally friendly cities.

One of our first conversations centered on what exactly it meant to be the world's greenest city. Is it the city with the most trees? The lowest greenhouse gas emissions? The smallest ecological footprint? Should it be a competition like figure skating, with marks awarded for both technical merit and aesthetic factors? Or should the contest be more like a decathlon, with cities earning points based on their performance in measurable categories such as water quality, area of green space, and public transportation? Gregor and I decided on the decathlon metaphor, to see how Vancouver stacked up against the world's greenest cities in ten key environmental categories.

Giant German engineering firm Siemens helped us by publishing comprehensive continent-by-continent assessments of urban environmental performance. Siemens used a decathlon-style approach as well, assigning cities points on roughly 30 indicators in nine categories: air quality, water, waste, land use, environmental governance, energy, buildings, transport, and climate change. The top-rated cities in Europe were the best in the world, led by Copenhagen, Stockholm, and Oslo. Cue the Danish national anthem. The top three cities in North America were San Francisco, Vancouver, and New York. In Asia and Latin America, the highest rated cities were Singapore and Curitiba. No African countries earned a high ranking, although Accra, Capetown, Casablanca, Durban, Johannesburg, and Tunis rated above average. The greenest cities in the world performed in a manner similar to the best decathletes, who have one or two events that are their specialty but achieve excellence across the whole suite of ten sports. When he set the decathlon world record at the American Olympic trials in 2012, Ashton Eaton finished first in seven of the ten events held over two days, while still ranking in the top eight in the other three events. In their Greenest City Index, Siemens noted, "Copenhagen's standout attribute is consistency," as it finished in the top five European cities in almost every category.

From a North American perspective, the results are both sobering and inspiring. Even the brightest lights in Canada and the U.S. are substantially behind the leading European cities in terms of measurable environmental performance. As we set about our work in developing a blueprint for Vancouver, we used the top European cities as models for targets, timelines, and tools that have proven successful. For example, many European cities have taken a leadership role in addressing climate change, going beyond Kyoto targets and pointing the way towards a future free from polluting and increasingly scarce fossil fuels. Key factors include how cities are designed, how urban infrastructure is built, and how people's day-to-day needs can be intelligently fulfilled in new ways. Whereas the average Canadian or American pumps out 15 tonnes of carbon dioxide annually through vehicle use, electricity consumption, and home heating/cooling, Vancouver residents generate the fewest per capita greenhouse gas emissions in North America, at 5 tonnes per person annually. Yet Stockholm residents have already reduced their carbon pollution to 3 tonnes per person, while the average resident of Oslo emits a mere 2.2 tonnes.

At the very first meeting of the Greenest City Action Team, a group of 18 business leaders, scientists, planning experts, and environmental activists led by Mayor Robertson, Councillor Andrea Reimer, and me, we brainstormed about what Vancouver could accomplish in a decade if it was genuinely committed to becoming the greenest city on Earth. There was palpable excitement in the air as people described their visions of the city's future. As if to endorse the ideas being discussed, a peregrine falcon swept down from the sky and landed on a railing just outside the window of our conference room. The meeting came to an abrupt halt as people rushed to get a photograph of this beautiful bird that has rebounded from the brink of extinction. It was, we all agreed, an auspicious sign, the embodiment of Vancouver's greenest city aspirations: to be the best in the world; to be a source of inspiration, optimism, and hope; and to demonstrate that concerted efforts can turn back the tide of ecological damage that humans have inflicted on the natural world.

Our team produced two plans. The first was a set of 48 short-term actions that could be started prior to hosting the Winter Olympics in 2010 to demonstrate that the city was serious about its environmental aspirations. To Vancouver's credit, 90 percent of those recommendations were either completed or well underway when the torch was lit at the Games' opening ceremonies. The second report laid out a series of ambitious targets, timelines, and actions in ten key areas including green jobs, climate change, buildings, transportation, access to Nature, waste management, air quality, water quality, local food, and ecological footprints. For example, Vancouver set a long-term goal of eliminating dependence on fossil fuels, with a short-term target of reducing greenhouse gas emissions to 33 percent below 2007 levels by 2020. To achieve these objectives, we proposed a wide array of actions, including increasing density, enabling mixed-use neighbourhoods, improving public transit, implementing a world-leading building code, investing in cycling and pedestrian infrastructure, and showcasing renewable energy at city-owned facilities. Vancouver City Council formally adopted the plan, and staff began the arduous task of developing detailed policies and programs to ensure the goals would be achieved. By 2013, Vancouver was already winning global recognition for its environmental progress, including the World Wildlife Fund's Green City Challenge Award and the World Green Building Council's Best Green Building Policy. Vancouver residents enjoy the best urban air and water quality in North America and have the lowest per capita carbon dioxide emissions, down almost 20 percent since 1990. *Time* magazine identified Vancouver as the healthiest city in the world. Like a champion decathlete, Vancouver is now in the top ten in almost every category evaluated in the Siemens Green City Index.

The city is changing in both visible and invisible ways. Separated bike lanes are being created along major roads. Community gardens, community orchards, and farmers' markets have sprung up in new locations. Like leading European cities, Vancouver is developing district heating systems. David Cadman is a former Vancouver city councillor (and president of the global organization

ICLEI—Local Governments for Sustainability) who was on our action team. After Vancouver won the right to host the 2010 Winter Olympics, David suggested using waste heat from sewage pipes beneath the ground to heat the athletes' village (which was subsequently converted into condos and social housing). It is a common practice in Europe but unprecedented in North America. When the city's engineer rejected the idea, Cadman said, "Here's the business card of an engineer in Switzerland, where they've been doing this for years." The Canadian engineer contacted the Swiss engineer, was convinced of the technology's merits, and eventually the sewage heat recovery system was built. It now supplies more than 70 percent of the annual energy demand in that neighbourhood, reducing greenhouse gas emissions by a comparable amount.

Using what were previously considered wastes as resources represents a brilliant shift in thinking that is new to North American cities but commonplace in Europe. European communities view waste as a valuable resource that can provide recyclable materials, compostable organic matter, heat for buildings, biofuels, and electricity. Carolien Gehrels of Amsterdam's city council argues, "Garbage is gold. Waste is not a problem but a valuable raw material." A state-of-the-art waste-to-energy plant in Amsterdam provides electricity to more than three-quarters of the city's households, while sending just 1 percent of the original waste to landfill. With more than 30 process innovations, the facility also enables recovery of valuable metals, from silver and gold to iron. Stockholm began building district-heating systems in the late 1950s. Today, their wastewater treatment system provides heat for over 100,000 homes (even on the darkest, coldest Swedish winter nights). Wastewater sludge produces a biofuel used by the municipal bus fleet. Cold water from lakes and oceans adjacent to the city is used for district cooling, a technology also used in Toronto. More controversially, Stockholm and other European cities burn garbage to generate both electricity and heat. Waste-to-energy plants produce 9 percent of Sweden's electricity and less than 1 percent of the waste created by Swedes goes to landfill. In recent years, a garbage shortage caused by effective reduce-reuse-recycle programs led Sweden to import trash from Norway to keep the waste-to-energy facilities in business.

Critics warn that burning garbage has three major downsides—hazardous air pollution, greenhouse gas emissions, and a disincentive to reduce, reuse, and recycle. Sweden addresses the first challenge with strict regulations governing emissions, requiring operators of these facilities to install state-of-the-art pollution abatement technology. Greenhouse gas emissions from electricity generated by waste-to-energy plants are higher than electricity from renewables but lower than coal or natural gas plants. The third critique is harder to rebut, although Swedes do generate less trash than Canadians or Americans and recycle comparable percentages of their garbage. In a circular economy, there would be no waste to burn in incinerators, because all products would be designed for reuse or recycling. Europe is now moving in this direction, recognizing that the various elements of the waste stream offer greater value if separated, recycled, and reused.

Following in the footsteps of Toronto and many other cities, Vancouver instituted a food scraps collection program as part of its efforts to reduce the amount of solid waste going to landfill by 40 percent by 2020, a milestone en route to the long-term goal of zero waste. At the same time, garbage pickup schedules were changed from weekly to bi-weekly, creating an additional incentive to reduce. The food scraps are used to produce compost that is used by the city's landscaping department and sold at garden centres. This initiative boosted Vancouver's diversion rate above 70 percent, referring to the proportion of solid waste that is either recycled or composted. Among the world leaders in diverting waste from landfill are San Francisco and Leipzig, Germany, both of whom have surpassed 80 percent. Getting to 100 percent is difficult but not impossible, and this goal has been formally adopted by hundreds of communities around the world.

North Americans use twice as much water every day as citizens of Europe, Asia, and Latin America. Residents of Amsterdam are the most water conscious in Europe, using only 146 litres per day, roughly half of New York, the most water-wise city in North America (262 litres per capita per day). All homes and businesses in Europe have water meters that measure use, and prices are two to three times higher than in Canada or the U.S. in an effort to use full cost accounting. The efficient use of water means lower costs, smaller treatment and wastewater infrastructure, and more water for wildlife and ecosystems.

It is vitally important to human health and well-being to be able to spend time in natural spaces such as parks, gardens, beaches, and forests. As the population density of cities increases (the inevitable consequence of growing populations living on finite areas of land), so does the value of protecting green and blue spaces. Almost every person in leading green cities like Vancouver and Stockholm lives within 300 metres of a park or other public green space. In Stockholm, public green space comprises a remarkable

40 percent of the city. The waterways in the centre of the city are so clean that you can swim in them, fish in them, and even eat the salmon and sea trout that you catch. When I visited Stockholm in 2007, I was amazed to see people fishing from the bridges adjacent to Sweden's Parliament buildings.

Trees are a valuable urban asset, providing an array of benefits that include cleaner air, lower temperatures, stormwater absorption, reduced erosion, wildlife habitat, improved water quality, lower energy costs, safer streets, and beauty. As they age, trees increase in value, making urban reforestation a great investment. A study done for Los Angeles indicated that, over a 25-year period, the health and economic benefits of planting a million trees would range from $1.6 billion to $1.9 billion. L.A. and New York each have initiatives underway to plant a million trees. Experts in urban forest management estimate that the tree canopy should cover about 40 percent of a city. World-renowned urban forests such as Vancouver's Stanley Park and Halifax's Point Pleasant Park offer irreplaceable reminders of the woods that once blanketed the land now occupied by Canadian cities.

JUST A CENTURY AGO, horses outnumbered cars on the streets of the world's major cities, causing a different kind of environmental problem. And just as the beginning of the 20th century witnessed a transportation revolution involving the end of equine dominance, the beginning of the 21st century is witnessing the decline of motor vehicles with internal combustion engines. After 100 years of planning and building North American cities around cars, walking, cycling, and public transit are becoming the preferred transportation options. The cities of Europe have always been ahead in this regard, and today two-thirds of all trips taken are on foot, by bicycle, or on public transit. In the greenest European cities, it's nine out of ten trips. In contrast, three out of four Canadians and Americans not only drive their own vehicle to work but do so alone. Walking, biking, and public transit account for fewer than one in ten American commutes and two in ten in Canada. In Europe, people dominate the streets, not motor vehicles. There is a critical mass of vibrant activity, ranging from sidewalk cafes to public art. Going for a walk in Paris, London, or Copenhagen is a completely different experience.

While some will argue that these benefits arise because European cities are much older, more densely populated, and have the built-in advantage of narrow streets, this is only a partial explanation. Amsterdam is a cyclist's paradise today but cars were predominant in the 1960s. Residents, galvanized by a growing number of traffic deaths and a proposed six-lane highway through the city's centre, joined forces to advocate for a different vision of the future in the early 1970s. The freeway plan was rejected and Amsterdam began prioritizing people over cars, which are now widely perceived as polluting, noisy, dangerous, and expensive.

The bicycle is the most energy efficient form of transportation, so a critical step in greening a city is to make cycling safe and popular, usually by creating separated bike lanes. These are not the bike paths common in North America, which are fine for recreation but often useless for commuting, visiting friends, or doing errands. Separated bike lanes create physical barriers between cyclists and drivers. This infrastructure is common in European cities and is becoming more widespread in Canada and the U.S. In Amsterdam, Copenhagen, and Groningen, more than half of all trips are made using bicycles, compared to 1 percent for the U.S. or 2 percent for Canada. Driven by public demand, bike-friendly infrastructure is on the rise. The proportion of people commuting by bicycle in some U.S. cities—including Washington and New York—doubled between 2009 and 2013. The total number of bike commuters grew faster than any other mode of transport in the U.S. between 2000 and 2012.

Vancouver has installed separated bike lanes on six major roads since 2010, in each case removing one lane from vehicles. John Neate Jr., owner of the JJ Bean cafes, happily agreed to replace two car parking spaces with bike racks, saying, "I think it's a super-smart move, just in terms of the number of bikes you can get in a car parking spot. We can have up to 20 bikes there." Even businesspeople who opposed the bike lanes have reversed their position. For example, restaurant owner Steve Da Cruz originally fought against the Union Street bike lane but now supports it, saying, "We have definitely benefitted from the increased usage of the bike lane." The total number of trips taken by bicycle in Vancouver is up 50 percent since 2008, and more than a million trips a year are being made on the separated bike lane crossing the Burrard Bridge.

Walkability is another essential feature of a green city. With proper urban design, people can live in complete communities, with a mix of housing, employers, schools, childcare, green spaces, shopping, and recreational opportunities all within a ten-minute walk. Older neighbourhoods in many cities often already enjoy these benefits. For most people, cars would rarely be needed. To encourage walking, Vancouver has been increasing the width of sidewalks; installing street furniture, maps, lights, water fountains, art, and other amenities; and making streets safer by lowering speed limits and using traffic calming. Walking now represents almost one in five trips in the city.

Stockholm demonstrates how cycling, walking, and public transit can be seamlessly integrated. When visiting Sweden, I wanted to go for a hike in a national park, despite the fact that it was October and there was a chill in the air. I took a bus to the city's central transit hub, then a train, then another bus. The price was cheap, each leg of the journey was comfortable, and the waits in between transfers were minimal. At each stage, I recognized a growing number of kindred spirits—Swedes wearing Gore-Tex and fleece and carrying daypacks. Only one in twelve Stockholm residents uses a private vehicle for commuting to work. Almost seven in ten people walk or cycle to work and school, even though the city has one of the coldest average temperatures in Europe. Cycling levels are up more than 75 percent since 1998, including an even more dramatic surge in winter cycling, which has more than doubled since 2005. One contributing factor is that bike lanes are now a priority when it comes to clearing snow. I was impressed by the routine availability of free compressed air stations throughout the city, enabling cyclists to inflate their tires whenever and wherever required.

All buses in Stockholm's city centre run on renewable energy. Stockholm has 400 ethanol buses, more than 100 biogas buses, and an electric metro system powered by wind and hydroelectricity. Stefan Wallin, environmental manager for Stockholm's public transit company, observed, "Regardless of political colour, politicians are quite keen to put their mark on sustainable development. We are continuously pushed to come up with new solutions."

Stockholm has piloted the use of at least ten different types of cleaner, greener buses, making it what Wallin described as "almost an engineer's playground." Another tool employed in Stockholm to encourage the adoption of greener transportation choices is a congestion tax. The tax was set up on a trial basis in 2006. The tax is one or two Euros, depending on the time of day, and applies to all vehicles entering the city centre between 6 a.m. and 6.30 p.m. Prior to implementation, public opinion polls indicated widespread opposition to the tax. At the end of the test period, polls indicated a change of heart. Traffic volume declined by 20 percent. People noticed the changes and appreciated less traffic congestion, cleaner air, reduced noise levels, and safer roads.

As our daughter, Meredith, gets older, we're riding bicycles more often. In Grade 2, Meredith rode her bike to the school-bus stop every day, except for a couple of times when she decided she'd rather run, and a couple of days where snow made it wise to walk. It's only 2 kilometres round trip, but it's a lovely way to start and end the school day. Her enthusiasm spilled over and got the neighbours, Isla and Owen, riding their bikes to the bus stop on a daily basis too.

Vancouver recently achieved its 2020 goal of half of all trips being made on foot, by bicycle, or via public transit. When Margot and I lived in Vancouver, we didn't own a car. We walked, cycled, and occasionally took public transit. We did join the upstart Vancouver Car Co-op, a car-sharing enterprise that enables members to use a vehicle when needed. At that time, car-sharing organizations were a novelty. Today, both non-profit and for-profit organizations have sprung up to offer their members access to cars on an as-needed basis. Globally, there are almost two million members of car-sharing groups, a number that's expected to reach 12 million people by 2020. Many millennials are choosing not to own cars but still need to use a vehicle occasionally. Research indicates that for every car used in such a program, 12 to 14 fewer vehicles are purchased.

Like car sharing, bike-sharing programs make bicycles available to individuals for short-term use. The first urban bike-sharing initiative debuted in the French city of La Rochelle in 1974 and is still operating today. Copenhagen's program, which started in 1995, was the first to use specially designed bicycles whose parts did not work with other bikes (reducing the incentive for theft). Bike-sharing programs are offered in over 500 cities around the world, more than double the number in 2011. The bike-sharing program in Wuhan, China, uses almost 100,000 bicycles.

One particularly promising conclusion of the Siemens Greenest City rankings was that "some cities with a below-average income clearly outperform their peer cities with higher incomes." A noteworthy example is Curitiba, a Brazilian city recognized as an urban environmental pioneer. The rapid bus system created in Curitiba has been emulated in hundreds of cities around the world. Rapid buses enjoy dedicated lanes and priority at intersections, while fees are collected prior to boarding to reduce delays. Curitiba was also the first Latin American city to establish a pedestrian-only street. In part because of these initiatives, residents enjoy good air quality. Curitiba also has a strong recycling program. Although I have not visited Curitiba, I have been to Bogotá, Colombia, where successive green mayors borrowed ideas from Curitiba and successfully adapted them.

CITIES MUST BE VIEWED as integrated systems where waste is a resource, density enables effective district heating and public transportation, and green spaces provide sanctuary for both people and Nature. The actions needed to make cities greener are good investments, paying dividends through healthier people, a more resilient economy, and a flourishing environment. The air will be cleaner, homes more comfortable, and getting around town more convenient. Leading green cities from Vancouver to Stockholm have demonstrated that going green provides not only the obvious environmental advantages, but also a host of additional economic and social benefits, ranging from the ability to attract talented immigrants and investment capital to lower healthcare costs and happier citizens. Vancouver and Stockholm both aim to be fossil-fuel-free by 2050 and are well on their way. The global competition for the title of Greenest City is motivating many cities around the world, from leaders to laggards, to revisit their land-use bylaws, transportation plans, and building codes to emulate the best.

Vancouver mayor Gregor Robertson's opponents have criticized him for going too far with his green initiatives. Separated bike lanes provoked a major backlash from some drivers, who accused him of waging war on cars. He was criticized as a puppet for developers because of actions to increase density along major public transit corridors. Yet the people of Vancouver clearly support the city's environmental leadership, as Robertson won a third consecutive term as mayor in 2014 after an election where environmental issues were at the top of the agenda. If he serves the full term, Robertson will become the longest-serving mayor in Vancouver history.

> *"Optimism is a strategy for making a better future.*
> *Because unless you believe that the future can be better,*
> *you are unlikely to step up and take responsibility for making it so."*
> —Noam Chomsky

Chapter 9
The Future of Buildings

I'VE LIVED IN ALL KINDS OF BUILDINGS over the past 50 years. In London, England, where I was born, my family lived in a small flat in a row housing development in Worcestershire Park. When I was two, we moved to Calgary, where I grew up in a 900-square-foot suburban bungalow. Despite the vicious winters, the house was framed with 2x4s, meaning the walls were only four inches thick. This was standard construction in most of North America until very recently. As a university student, I lived in cheaply built high-rise buildings in Edmonton and Montreal. During law school in Toronto, my homes included several cockroach-infested rooming houses and a brick duplex. In Guatemala, I shared a single-storey, two-bedroom cinderblock home with nine members of an extended Mayan family. As a young lawyer in Vancouver, I lived on the top floor of a friend's house. Although Vancouver's climate is relatively mild by Canadian standards, the wind still had a bite in winter months, blasting through and around the single-pane wooden windows. And finally, in our first foray into home ownership, Margot and I bought a gorgeous but rundown house on Pender Island, nestled between Vancouver and Victoria on Canada's west coast (where we still live). Although renters had used the house as a marijuana grow-op, it was clear that the original architect and builder put substantial thought into conserving energy. The house faces south, where large windows capture the heat and light of the sun. The walls are six inches thick and the windows are double pane. Good, but not great. Better than all of my previous homes, which were in colder climates but had thin walls and single-pane windows and were not designed to take advantage of the sun's light and warmth.

For decades, the overwhelming majority of buildings in wealthy countries have been constructed based on the assumptions that energy would always be cheap and its use devoid of meaningful consequences. Today, those assumptions are no longer tenable. The OPEC oil embargos of the 1970s and the ensuing price shocks marked the first time that energy conservation and energy efficiency became topics of everyday conversation. Yet only a tiny fraction of homebuilders and homebuyers took the warnings seriously. Visionaries across the world experimented with new ideas for designing and building shelter. Inevitably some experiments failed—aesthetically, functionally, or economically. The first generation of energy-efficient buildings were built airtight to keep the heat in and the cold out, or vice versa during the summertime. Many of these well-intentioned structures created terrible moisture problems and burdened occupants with a degree of air quality that ranged from unpleasant to unhealthy. Other experiments were successful and have been replicated, revised, and repeated.

As a result, today we're in the midst of an extraordinary—yet largely invisible—transformation of the buildings in which we live, work, study, and play. Although oil and gas prices may swing up and down, the previous assumption of cheap, limitless energy has been replaced. We are belatedly but painfully aware of the impacts of burning fossil fuels on human health, ecosystems, and the climate.

Incredible design and technological advances have ushered in a new generation of buildings—residential, commercial, and industrial—with either minimal environmental impacts or net benefits. No longer just utopian fantasies in glossy architectural magazines, buildings that require little or no heating or cooling, are self-sufficient in water use, and generate more electricity than they use are en route to becoming the new norm. For the first time in decades, the average size of new homes in North America is declining. At the same time, existing buildings are being extensively renovated to reduce their ecological footprints.

As I was doing research for this book, I read a magazine article about a different kind of house being built in Victoria, the first of its kind in that city. Mark Bernhardt and his father were planning to meet a set of standards created in Germany for what is known as a *passivhaus*, or passive house. The exacting German construction standards cut energy use for heating and cooling by as much as 90 percent. I wanted to see the future of energy-efficient building for myself, and the Bernhardts were accommodating. Nestled in the Cedar Hill community, the house stands out visually from its neighbours mainly because it is newer. From the outside, it's an attractive residence similar to what you might find in the pages of *Dwell* magazine, with sharp modernist angles and an eye-pleasing blend of wood and glass. The Bernhardt home is a duplex, with Mark's parents living on the ground floor and Mark's young family upstairs. Each level is 1,900 square feet, making it large compared to European homes but relatively modest by today's super-sized North American standards.

The first thing you notice when you step into the Bernhardt's family home is the quiet. The front door closes with a satisfying click, like the door of a bank vault. Traffic noise and the background hum of the city disappear. The interior design is generic, attractive but not particularly memorable. Huge windows cover much of the south and west facing walls to welcome the sun's heat and light during fall, winter, and spring. Overhangs have been carefully designed to block the sun when it's high in the sky during summer, keeping the house cool. I visited in February, on a steely grey day when the outside temperature was around 8°C. Inside, the temperature was an even 19°C, yet the heat was not on. There was no fireplace. Sunshine plus the heat from appliances, lights, and human bodies usually provides sufficient warmth. After an hour in the living room discussing the various features of the house, I was struck by how comfortable it felt. The room was warm without overheating. Even on a blustery winter day, there was not the slightest hint of a draft. And as a *New York Times* article on a passive house said, "The air inside the house feels so fresh, you can almost taste its sweetness."

The critical differences that set the Bernhardt's house apart are largely invisible at first glance. The walls are one foot thick, with an eight-inch outer layer and a four-inch inner wall where all of the wiring, air ducts, and plumbing are contained. Great care was taken during construction to eliminate all unnecessary openings in the outer wall, maximizing its insulating effect. Regular insulation was used to fill the wall cavities. Triple-pane windows let the sun's energy in, but not out. Because passive houses are so airtight, they incorporate a heat recovery ventilator, which exchanges stale indoor air with fresh outside air, transferring the heat from outgoing air to cold incoming air in the winter (and vice versa in the summer). A very small central heating unit can use the ventilation system to circulate heat on the rare occasions when it is needed. The windows can be opened to boost air circulation.

The Bernhardts kept a very close eye on their building costs. Their house was actually less expensive than a comparable custom built home. That's right—this comfortable, attractive, super energy-efficient home cost less. The premium windows, thicker walls, extra insulation, and additional design costs were partially offset by major savings on the much smaller heating system. Monthly mortgage payments are larger but are more than offset by lower utility bills. Although their first winter in the house was colder than the average Victoria winter, heating costs were even lower than anticipated. This was the first passive house that Mark Bernhardt has built, and future buildings will only get better. As Mark said, "Our residence includes huge windows, is not a particularly compact design, and does not have great solar exposure, but it still works."

According to data compiled by Natural Resources Canada and the Canadian Mortgage and Housing Corporation, the average Canadian home uses 170 kWh of energy per square metre per year for heating and cooling. I know that seems abstract, but please bear with me. New homes, built with the benefit of slightly stronger building codes, use 150 kWh/m^2/year. An Energy Star home, built to meet voluntary energy efficiency guidelines, uses 100 kWh/m²/year. The standard required to have a building certified as a passive house is a maximum of 15 kWh/m²/year. In other words, a passive house uses 90 percent less energy for heating and cooling than a regular, newly constructed Canadian home. A 90 percent reduction in energy use takes a tremendous amount of pressure off the environment—reducing air pollution, greenhouse gas emissions, and harm to biodiversity, while maximizing the homeowner's savings.

ONCE A BUILDING'S ENERGY USE has been reduced so dramatically, it's easy to take the next step. A passive house uses so little energy that with the addition of a small solar photovoltaic system, it can become a net-zero energy house, producing all of the electricity that its residents use over the course of a year. A widely adopted policy called net metering even allows building owners who produce more electricity than they consume to sell the surplus to their utility company. Imagine getting a cheque from your electric utility instead of a bill! In Canada, one of the first net-zero homes was constructed by green builder Peter Amerongen as a pilot project. A decade ago, Amerongen and 45 energy and building professionals collaborated on an experimental net-zero home project in Edmonton, Alberta, where winter temperatures dip as low as −40°C for weeks at a time. The designers and architects focused on three basic steps. First, maximize free energy from the sun through passive solar design. Second, minimize energy consumption through the use of smart, energy-efficient technologies. Third, harvest renewable energy, generally by installing a solar PV system. Conceived in 2004 and completed in 2007, the Riverdale project has served as an inspiration to many architects, designers, and builders.

A leading Alberta homebuilder, Landmark Homes, is pledging that every home it builds will now be net-zero. The company constructs more than 800 homes a year, using a state-of-the-art factory that prebuilds walls to exacting specifications, creating higher quality homes at lower cost. According to Reza Nasseri, Landmark's CEO, they decided to build net-zero homes "because we believe very strongly in the future of our children and sustainability is probably the most important thing today in our lives and we really have to protect the Earth for our new generation." Landmark recently completed the Sparrow Landing townhouse development in Edmonton, a project that includes 14 net-zero homes. Standard features include airtight construction, active heat recovery ventilation systems, triple-pane windows, solar PV systems, ground source heat pumps, Energy Star appliances, low-flow toilets, high-efficiency furnaces and water heaters, and a drain-water heat recovery unit that recaptures heat from used hot water. The additional costs per unit are roughly $35,000 per 1,700-square-foot home, offset by the absence of natural gas bills and minimal electricity bills. Thanks to the dramatic decline in the cost of solar panels, building a net-zero home has gone from a radical pilot project in 2004 to commercially viable in just ten years.

The burning question is: Why aren't all homes built to this standard in countries like Canada and the United States? Super-efficient buildings have a strong economic case, deliver multiple environmental benefits, and, most importantly, are fabulous to live in. Growing recognition of these facts led a number of European cities to incorporate passive house standards into their building codes. The success of these early adopters pushed Europe to the cusp of a revolution in construction practices that will make passive houses and similar super-efficient buildings the norm rather than the exception. The European Union's new building directive requires all commercial buildings as of 2020 and all residential buildings as of 2021 to be near-zero energy buildings. In effect, all new buildings in Europe, in about five years, will be as energy efficient as passive houses—ten times as efficient as average new North American buildings.

California is emulating Europe and requires net-zero energy construction for all new homes by 2020 and all commercial buildings by 2030. Like the Riverdale house in Edmonton, new buildings in California will emphasize passive solar design, maximize energy conservation and efficiency, and produce electricity using solar panels. California will also require existing buildings to move towards net-zero when owners undertake remodelling, renovation, or repurposing projects that exceed certain thresholds. Governor Jerry Brown ordered state agencies to achieve net-zero for half of the square footage of existing state-owned buildings by 2025. Research conducted for California shows that net-zero energy construction will add $4,000 to $16,000 to the cost of an average home. Considering that utility bills will be almost eliminated, this is an excellent investment.

AT THE CUTTING EDGE of eco-friendly construction today, even beyond net-zero energy, is a concept called the living building. At the heart of the VanDusen Gardens, a 22-hectare urban oasis in Vancouver, the new visitor centre is one of the most beautiful buildings I've ever set foot in, simultaneously airy and warm.

Architect Jim Huffman, who designed the building with colleagues Peter Busby and Cornelia Hahn Oberlander, said that the team was inspired by a native orchid. According to Huffman, the design "considers the complete system of environmental technologies—how the building operates, what it's made of, and the nature and ecological systems around it, and tries to create a situation where the building becomes a positive factor in all of those things." Undulating roof segments, like the petals of a flower, crown thick walls made of rammed earth and concrete. To the east are solid walls that dampen the noise from busy Oak Street, while to the north and west, glass walls face the gardens and ponds. Within the 20,000-square-foot building are a cafe, library, art gallery, offices, volunteer facilities, garden store, classrooms, and rental spaces. The visitor centre is the first project in Canada to register for the Living Building Challenge, a super-stringent system of measuring sustainability that goes substantially beyond passive house standards and the better-known LEED certification (Leadership in Energy and Environmental Design). Living buildings must be net-zero energy, net-zero water, non-toxic, beautiful, local, and educational. All interior spaces must have direct access to fresh air and daylight. For every hectare of land disturbed, a hectare of wildlife habitat must be protected. To date, only a handful of buildings in the world have earned full certification as Living Buildings. The first project in Ontario to pursue this status is York Region's Forest Stewardship and Education Centre, completed in 2014, while the Mosaic Centre, a 30,000-square-foot building in Edmonton, will be Canada's first office building to meet this standard. The Mosaic Centre harnesses geothermal and solar energy, was completed ahead of schedule, and came in under budget. In the words of co-owner Christie Cuku, "We said it was going to be beautiful. Check. Sustainable. Check. And affordable. Check."

The Living Building Certification process requires the use of local materials, non-toxic construction products (eliminating the use of common products such as PVC piping and pressed wood products containing formaldehyde), a minimum threshold of recycled materials, and the use of only reclaimed wood and wood certified by the Forest Stewardship Council. The VanDusen visitor centre used salvaged wood, trees that had naturally fallen at the end of their lifespan, and a few trees cut down on-site. In terms of performance, living buildings must produce at least as much water and energy as they consume. Rainwater at VanDusen is captured by a green roof planted with native species. The water is filtered and stored in a large cistern. Wastewater is treated and discharged on-site. Energy is produced by solar panels and a geothermal system, and a solar hot water system contributes to heating and cooling. The green roof helps insulate the building while an operable glazed oculus assists with air circulation to keep the atrium cool. Wide roof overhangs prevent solar heat gain during summer months while enabling it in cooler months. Architect Jim Huffman said that the toughest part of the Living Building Challenge "is the Materials Imperative, which calls for avoiding items on the Red List, a list of substances determined to be detrimental to human health and the environment."

Another Living Building recently finished in Vancouver is the UniverCity Childcare. Heralded as the greenest childcare facility in the world, the building actually cost 18 percent less than conventionally constructed daycares built in the region at the same time. Opened in 2012, with room for up to 50 children aged three to five, it's a lovely building. Children were actively involved in the design process, contributing fresh ideas that were incorporated into the facility. Dale Mikkelsen, director of development for the project, said, "What if these kids could spend three years of their lives resetting their normal—expecting that we treat and reuse our rainwater, expecting that we generate the energy we need, expecting that you can eat the building and not get sick, expecting that the building will still be there for their children and their grandchildren?" The outside areas are intended to stimulate kids' imaginations, enabling them to explore air, water, light, gravity, vegetation, and the changing of the seasons. When I visited on a sunny summer day, all of the children were outside laughing, playing games, ducking in and out of woven willow huts, and building sand castles. It looked like the kind of place any kid would love to spend their days. As a parent, I would not hesitate to send my child to such a thoughtfully built facility. It's as green as buildings get, it's beautiful and optimally equipped to meet the needs of the building's users, it will result in healthier children, and it cost less than conventional construction. This is the future of shelter.

IT'S MORE CHALLENGING TO RENOVATE existing buildings than to build new ones to high environmental standards. However, a few relatively simple steps can be taken to cut the energy use of existing buildings by one-half to two-thirds. The most effective actions will vary from region to region and building to building, but there is one general imperative: minimize the energy used for heating and cooling. The first step is to get a professional energy audit, which will focus on four key areas: airtightness, insulation, heating/cooling systems, and windows.

At our Pender Island home, we commissioned an energy audit and have taken a number of recommended steps, including weather-proofing and draft-proofing, turning down the temperature on our water heater, adding insulation in key areas of the house, and replacing some windows with more energy-efficient models. We use electricity for heat during the winter, so we also installed a high efficiency wood stove to reduce our electricity costs. With reduced electricity consumption, the falling cost of solar, and a south facing roof, we are now planning to install a solar PV system that will enable us to become a net-zero home, capable of selling solar power back to B.C. Hydro. Solar PV panels and solar hot water heaters are good investments today, and will only get better as their prices fall and electricity and natural gas prices climb. My younger brother Sandy is putting the green building principles discussed in this chapter into action in building his family's dream home on the outskirts of Nelson, B.C. The gorgeous off-the-grid home features passive solar design, a Finnish-inspired masonry heater, walls that are 12 inches thick, and electricity generated by a small creek nearby.

AS THE BERNHARDTS' PASSIVE HOUSE and the UniverCity Childcare prove, today's green buildings provide superior comfort and performance at a cost that is equal to or even less than conventional building. Many studies have shown that in green buildings students perform better on exams, workers are more productive, and all occupants are healthier and happier. When all of the broader societal benefits are considered, including the improved health and well-being of building users, the reduced energy, water, and material footprints, and the associated environmental benefits, there's an irrefutable case that all jurisdictions should emulate Europe and California by enacting stronger building codes that require super green buildings.

"Between the optimist and the pessimist, the difference is droll. The optimist sees the doughnut; the pessimist the hole!"
—Oscar Wilde

Chapter 10
Electrifying Transport

SINCE THE TURN OF THE CENTURY, cracks have begun to appear in society's century-long love affair with the car. For the first time in history, people in many wealthy countries are driving less. The former symbol of freedom and autonomy is increasingly viewed, especially by young people, as more of a burden than a blessing. Fewer young adults have driver's licences in Canada, the U.S., the U.K., Germany, Japan, Sweden, Norway, and South Korea. The portion of Americans aged 16 to 24 who have driver's licences fell to its lowest level in more than 50 years. Vehicle ownership is also down substantially, and even those who drive are logging substantially fewer kilometres behind the wheel each year. For example, from 2001 to 2009, the number of "vehicle miles" travelled by all Americans declined. The drop was most pronounced among individuals aged 16 to 30, whose mileage dropped 25 percent. At the other end of the spectrum, the population is aging and seniors generally drive less than other adults. Rates of cycling, walking, and taking public transit have all spiked, generating demand for improved infrastructure—wider sidewalks, separated bike lanes, and more buses and trains. "We're at the beginning of the end of the age of the car," said Vancouver city councillor Geoff Meggs. In Japan, there is a newly minted expression *kuruma banare*, which roughly translates as "walking away from the car."

These trends are particularly pronounced in cities, where increasing density makes owning a car unnecessary or even problematic. New York City has been allocating lanes away from cars to make room for cyclists, pedestrians, and the kinds of car-free public plazas that contribute to the vibrant culture of many European cities. In New York, Washington, Boston, Philadelphia, San Francisco, and Baltimore, more than 30 percent of households are carless. In Toronto and Vancouver, the proportion of trips made by car has been falling since the 1990s. Cities have relaxed rules that previously required developers to build at least one parking space for every unit in their buildings.

This may be one of the few times in my life when I was actually ahead of a trend, as I did not get my driver's licence until I was 25. Like today's millennial generation, I travelled on foot, by bike, and public transit despite the frigid winters in Calgary, Edmonton, and Montreal. Today's youth have diverse motivations including environmental concerns, the high cost of car ownership, and access to the Internet. According to Maurie Cohen, a social scientist at the New Jersey Institute of Technology, "Teenagers and 20-somethings have basically traded in a two-and-a-half-ton vehicle consuming large volumes of fossil fuels for a device that fits into the palm of their hand as a means of maintaining social connectivity."

Obviously I've never been a real "car guy." I didn't own a car until becoming a father at the age of 41. So it's kind of funny, in retrospect, that one of my first job interviews after I earned a business degree from the University of Alberta was with one of the Big 3 American automakers. It was for some kind of marketing position, but when one of the icebreaker questions was "What do you drive?" things went off the rails. When I confessed having taken a bus to the interview, the man literally gasped. Needless to say, I wasn't offered the job.

Two decades flew past before I bought my first car, which was a 2006 Toyota Prius. It's still our car. Margot and I bought it for three reasons. First, Meredith was about to be born and we were planning to live on Pender Island in a rural community. Second, it was the most fuel-efficient vehicle available in Canada at the time (roughly 50 miles per galon or 5 litres per 100 kilometres). The clincher was that every taxi driver we'd spoken to in the preceding years raved about the car. I once met Andrew Grant, the Vancouver cabbie who racked up so much mileage on his first-generation 2001 Prius that Toyota bought it from him to see what, if anything, was wearing out. We looked into getting a used Prius, but at that time Toyota was only making hybrids at one factory in the world. Because demand was so much greater than supply, it was actually less expensive to order a new one, although we had to wait three months for delivery.

At the time, critics were warning that Prius owners could expect to pay thousands of dollars for replacement battery packs every few years, but it turns out that Prius battery packs have rarely worn out. Cab driver Andrew Grant replaced his original Prius with a 2004 version that he used as a taxi for over 1.5 million kilometres before retiring it in 2011 with the original battery pack still in the car. From Vancouver to Vienna, Priuses used as cabs have racked up more than one million kilometres without requiring anything more than routine tune-ups, new tires, a few new parts, and refreshed upholstery. The overwhelming majority of the taxis in Vancouver and Victoria are gas-electric hybrids. In a recent taxi ride to the Victoria airport, I asked the driver how many kilometres were on his Prius. 800,000, he replied. Have you replaced the batteries? No. Have you replaced anything? Yes, the seats wore out.

Meredith was born in January 2006. In February, our Prius arrived. It was relatively anticlimatic. Mostly Margot drove it. It got us from A to B. In March, I went on a trip to Boston with David Suzuki to talk about our report *Sustainability Within a Generation* at a Harvard conference. When I returned to Vancouver, Margot picked me up at the airport en route to the ferry terminal and home. Meredith gurgled away happily in the back and I assumed my usual role as navigator. As we merged onto Highway 99, I noticed a red exclamation mark flashing on the dashboard and asked Margot about it. "Oh, that's been flashing for a while," she replied. I dug the owner's manual out of the glove compartment and began searching for information on the red light.

"Oh-oh," I said.

"What is it?" Margot asked.

"We're out of gas."

We were on the highway, zooming along at 90 kilometres per hour and had used up every drop in the gas tank. A Prius is a hybrid, meaning it has both a gasoline engine and an electric motor. Using the brakes recharges the battery pack, so the car never needs to be plugged in. But the car is not engineered to run solely under electric power. In a mild panic, we took the first exit, hoping to make it to a gas station. Instead we found ourselves surrounded by prime farmland, with the helpful dashboard computer screen showing us that the battery was as drained as the gas tank. We rolled to a stop in the middle of nowhere, out of gas and electricity. We had a good laugh and then Margot called Toyota's Roadside Assistance number. They promised to send someone to help us. A very nice woman eventually arrived in her tow truck and gave us a few litres of gasoline. She asked us to start the car to ensure it was working and we replied that it was already running. She looked at us like we were nuts, as she was clearly unfamiliar with the silence of the Prius. I still tease Margot about not knowing she needed to fill 'er up. We joked about routinely driving until we ran out of gas, then calling Toyota. The car has over 60,000 kilometres on the odometer and (touch wood) hasn't had any mechanical problems. We rarely take our car off the island except for longer trips, because Victoria and Vancouver both have decent transit systems and walkable downtowns.

ALTHOUGH WE'VE ENJOYED OUR PRIUS, it still burns gasoline, with all the negative impacts that produces. We've started looking into fully electric vehicles, as a growing number of manufacturers are now selling them, including the Nissan Leaf, BMW i3, Chevrolet Spark, Fiat 500e, Ford Focus, Smart Car, Mitsubishi i-MiEV, and Volkswagen e-Golf. Ford, Honda, Kia, and Mercedes also sell electric versions of some of their popular models. The major drawbacks of these vehicles are higher prices and limited range—the least expensive available in 2015 retails for roughly $25,000 (although government rebates may lower the price) and none will go further than 100 miles (160 km) before needing to be recharged.

The limited distance that electric cars can travel between charges led to the coining of a new phrase: range anxiety, the fear that a vehicle's batteries will run down before the destination is reached, leaving the occupants stranded. I borrowed a friend's Nissan Leaf for a day. It was quieter than our Prius, because there's no gas burning engine, but otherwise driving it felt quite similar. The Leaf has a range of 150 kilometres but may not reach that in cold weather or on hilly terrain. With government rebates, in 2015 you can take home a new Leaf in the U.S. for just north of $20,000. Used Leafs are even cheaper. That's quite a deal for a car that you'll never buy gasoline for. And people are grasping the benefits—sales of the Leaf set new records in the U.S. each month for 20 consecutive months in 2013–2014.

Maybe it's because I'm on the cusp of middle age, but I have to admit being mildly starstruck by the cars produced by Tesla Motors. You've likely heard the name, though it's less likely that you've seen their vehicles, and extremely unlikely that you've driven in one or own one (unless you live in Norway). Tesla is a California-based electric car manufacturer with just three models to date (though more lie, tantalizingly, on the horizon). There's the Roadster, a sporty little two-seater; the Model S, a luxury sedan; and the Model X, a sport utility vehicle. They look gorgeous, and best of all, they are fully electric zero-emission vehicles. They are the most expensive electric cars and offer by far the longest range. Of course, there are several caveats. All motor vehicles consume energy and materials in the production stage. All motor vehicles require roads, parking space, and other expensive, environmentally harmful infrastructure. All motor vehicles can contribute to the terrible toll of deaths and injuries inflicted by accidents. And even with zero-emission electric vehicles, their hue of green depends on the source of the electricity used to recharge the batteries. A coal-powered electric vehicle has a larger footprint than a solar or wind or hydro-powered version.

The Tesla Model S got the highest approval rating ever from *Consumer Reports*. Giving it a 99 out of 100, the magazine said the car's performance was "off the charts." Jake Fisher, head of vehicle testing for *Consumer Reports*, said, "We don't get all excited about many vehicles, and with this car we really did." The *New York Times* review said the Model S "is simultaneously stylish, efficient, roomy, crazy fast, high tech." It was the unanimous choice for *Motor Trend*'s 2013 Car of the Year award. It has the highest possible safety rating. The Model S can travel up to 480 kilometres on a single charge. Networks of high-speed charging stations are being built across Canada, the U.S., Europe, and China, enabling an 80 percent charge in just 20 minutes, lessening range anxiety. By 2014, Tesla owners could drive across the U.S., from California to New York, never having to pay a penny for fuel.

In researching this book, I test drove my friend Mike Cormack's Model S in Vancouver. The Tesla is clearly a high-performance driving machine. It's very sleek (to reduce wind resistance), the interior feels luxurious, and there are some cool high-tech features like recessed door handles that pop out as you approach the car. The really big difference is power—when you press the accelerator (no longer a "gas" pedal), the Tesla really surges forward. It can go from zero to 60 in less than five seconds, comparable to sports cars. Most of the buttons, levers, and knobs in a regular vehicle have been replaced by a large touch screen on the dashboard to the right of the driver. The 17-inch touch screen is like a giant iPad, controlling headlights, climate, navigation, entertainment, and access to the Internet. It even allows you to open the sunroof or see behind the car with a rearview camera. Voice control is optional. There is a ton of luggage room both under the hood and in the trunk. Driving around Vancouver on a rainy day, the ride was so quiet and so smooth that it was unlike any other trip I have ever taken in a motor vehicle. The Model S represents both the future of driving and a nail in the coffin of the internal combustion engine.

Despite the high price and limited range, in 2013 and 2014 the Tesla Model S and the Nissan Leaf sat at the top of the monthly sales charts in Norway. In other words, they were the top-selling vehicles in the whole country. Why are they so popular in Norway? The answer begins with acknowledging that Norwegians are a fairly wealthy people. There's also a pretty good network of EV charging stations. About 97 percent of Norway's electricity is produced by hydroelectricity, so electric cars run on green power. An important factor is that in 2014 gasoline was expensive at $2.40 per litre (almost $10/gallon). But the most important reason is likely the government's levy on new car purchases, which is based on vehicles' fuel consumption. A car that uses 12 litres per 100 kilometres (20 mpg) will incur a levy of $45,000. Yes, you read that correctly. In some cases, the levy can exceed the retail price of the vehicle. But since the Leaf and the Tesla consume no fuel, there is no levy. To a Canadian or American, the Norwegian motor vehicle levy may seem extreme. In fact, it is economically and environmentally sound, as it implements the polluter-pays principle, putting a dollar value on the health and environmental costs associated with driving and burning fossil fuels.

In 2014, Tesla made the astonishing announcement that all of their patents would be made publicly available in order to accelerate the spread of electric vehicles. CEO Elon Musk explained that Tesla itself could not possibly build electric vehicles fast enough to avert the global climate crisis. Musk said patents "serve merely to stifle progress, entrench the positions of giant corporations, and enrich those in the legal profession." This was music to the ears of Chinese billionaire Lu Guanqiu, chairman of Wanxiang Group, who made his money selling tractors but now wants to become a global leader in electric vehicles. He bought Fisker Automotive in a bankruptcy auction and plans to make his company Tesla's number-one competitor. "I'll put every cent that Wanxiang earns into making electric vehicles," he said. "I'll burn as much cash as it takes to succeed, or until Wanxiang goes bust."

Moving towards a transportation system based on renewable energy presents a stiff challenge. Electric vehicles (EVs) still make up a relatively tiny share of the market, but that share is growing exponentially. In 2000, there were fewer than one thousand fully electric vehicles sold in the entire world. By 2011, that figure had grown to 45,000, and by 2014 more than 400,000. Sales of electric vehicles are doubling each year, albeit from a small baseline. In Europe, this is the fastest growing segment of the motor vehicle market. Today, more EVs are sold each month in Norway than were sold globally a decade ago. In the Netherlands, where gas prices are similar to Norway, the EV market share increased 1,900 percent from 2012 to 2013. China is investing over $16 billion in electric vehicle charging infrastructure. Tesla has installed hundreds of charging stations in China, while Chinese EV companies operate even larger networks. By law, until 2016 at least 30 percent of the vehicles purchased by the Chinese government must be electric vehicles. That requirement is expected to increase after 2016, as China continues to grapple with the consequences of its pollution.

The growing success of EVs has been assisted by falling costs. Electric vehicle battery prices are down a staggering 500 percent in recent years. Tesla is building a $6.5 billion battery gigafactory in Nevada that is expected to produce innovations that drive prices down even further. BYD is building similar facilities in China. Disruptive leaps in storage technology are possible that could provide major boosts for EVs. As more countries enact policies like Norway and the Netherlands, the shift will accelerate. According to auto industry experts Navigant Research, less than half of the vehicles on the world's roads in 2035 will have internal combustion engines. The transformation is being driven by a combination of rising fuel costs, environmental concerns, and technological innovations. There is also a persistent rumour that Apple is working on an electric vehicle to be unveiled by 2020.

Imagine integrating a net-zero home with an electric vehicle. Soon electric vehicles will be part of the smart grid, an enhanced version of today's electricity distribution system. The battery systems in EVs will be able to store excess electricity and supply it back to the grid when needed. This will improve the grid's stability and reliability, enhance the usefulness of renewable energy sources, and provide a revenue stream for EV owners. As electric utilities move towards time-of-use pricing (charging higher electricity prices at times when demand is higher), EVs will be able to supply electricity to the grid at higher prices and recharge at lower prices. The key is to store energy produced by intermittent renewable sources like wind and solar until it is needed. Germany and Japan are once again in the lead. The German government is subsidizing residential solar PV systems that incorporate energy storage. Japan, in the aftermath of Fukushima, is expected to add 400 megawatts of solar plus storage by 2018.

ELECTRIC BICYCLES ARE PERHAPS not as sexy but are far more common and environmentally friendly than electric cars. E-bikes have small motors that assist in propulsion while the rider is still able to pedal. Usually powered by batteries, e-bikes range from those that offer a small boost in climbing hills to vehicles that are closer to mopeds. China has nearly 150 million e-bikes on the road. European sales of e-bikes are approaching a million bicycles per year, led by Germany and the Netherlands. Adoption in North America is slower, but growing.

Even greener than electric cars are electric trains. Chinese high-speed rail launched in 2007 and already leads the world with over 11,000 kilometres of track and over 1.3 million riders daily. That constitutes more high-speed rail than the rest of the world combined. Another 10,000 kilometres of track is under construction or planned in the next five years. Almost all of Japan's extensive rail network is electric, including all high-speed trains. In contrast, most of the trains in the U.S. still rely on diesel, and there is very little high-speed service. Canada is the only G8 country with no high-speed trains.

Ground-breaking electric ferries have now been introduced in Sweden and Norway. In Sweden, the *Movitz* carries 100 passengers on a one-hour run and recharges its battery while people get on and off the boat. Norway launched the ZeroCat 120 in 2015, an electric ferry with the capacity to carry 120 cars and plug into a supercharger when it docks. As battery technology improves, the shift to electric ferries will accelerate.

A number of companies make electric aircraft, though none of these vehicles has anywhere near the ability to displace regular airplanes (yet). In 2013, a solar-powered airplane flew across the U.S., although it required quite a few stops to get from California to New York. The future of these technologies, as well as ballyhooed prototypes of self-driving cars and solar roads, is unclear, but they are unlikely to make a major impact in the short term. Biofuels are controversial, but have a part to play in the transition away from fossil fuels, as long as used in moderation and produced with agricultural residues or fast-growing crops raised on marginal lands, so that food production and biodiversity are protected.

Hydrogen vehicles have long been mentioned in conversations about the future of transport. But there are hurdles that will be difficult to overcome, including high costs, the lack of a distribution network, and perhaps most damning, the fact that these vehicles are much less energy efficient than electric vehicles. For example, a BMW hydrogen vehicle prototype uses 250 kWh per kilometre while an electric vehicle uses 20. For this reason, Cambridge University energy professor David MacKay says simply that hydrogen vehicles are a "bad idea."

THINK ABOUT THE EVOLUTION of camera technology. I grew up in an era when every single camera used film. When digital cameras were first introduced, the majority of people were skeptical. Digital cameras were more expensive and there were many unanswered questions about the quality of the photographs. Yet within the span of about a decade, the market completely transformed. Prices of digital cameras dropped dramatically, their advantages became apparent, and the disadvantages largely disappeared. The same process seems likely to occur in the transition to electric vehicles. Every single person whom I've spoken to about their EV experience is wildly enthusiastic. Driving EVs is fun and easy, and people love never having to pay for gas again. Like early digital cameras, EVs today are more expensive, but lower operating costs and government subsidies offset much of the higher purchase price. As the technology continues to improve and prices continue to fall, EVs will gain a growing share of the market, relegating gas-powered vehicles to a niche similar to film cameras today.

"When I look at the future, it's so bright it burns my eyes!"
—Oprah Winfrey

Conclusion
From Optimism to Action

THE FRASER RIVER, which runs through Vancouver, is the greatest salmon river on the planet, the freshwater gateway to the vast interior forests of British Columbia and their myriad streams. Each year, millions of sockeye salmon suddenly curtail their oceanic wanderings across the vast Pacific Ocean. Driven by a relentless urge to return home, they navigate their way back to the estuary at the mouth of the mighty Fraser. There they undergo a stunning transmogrification, from silver bullets into crimson humpbacks with olive green heads and distended jaws. Sockeyes' amazing journey goes up the Fraser and across the Coast Mountains to spawn the next generation.

Every fourth year, a dominant sockeye run returns to the Adams River via the Fraser. It's an incredible spectacle that our family loves attending. Located about 400 kilometres northeast of Vancouver, the Adams River is sockeye spawning nirvana. The water is cold and clear, with a bed covered in the optimal type and size of gravel. It lies in the shadow of the mountains, with trees garnished in dazzling fall colours. The local community hosts a Salute to the Sockeye festival every fourth year, drawing tens of thousands of visitors. Every day during the peak of the run, dozens of yellow school buses roll into the parking lot bearing bright-eyed schoolchildren excited to see the spawning sockeye.

Meredith first visited the Adams River sockeye run in 2006 when she was nine months old. In 2010, as a five-year-old, she witnessed an estimated four million sockeye spawning. The smell of so many dead and decomposing fish is powerful but not revolting. Eagles, bears, otters, mink, ravens, and other creatures gather to share the bounty. Bears and humans co-exist according to unwritten rules about visiting hours. First thing in the morning, you'll often see paw prints and steaming piles of bright orange bear scat, as though someone emptied a dozen tins of canned salmon into a pile on the trail. Yet in all our visits, we've never seen a bear.

The Adams River sockeye survived decades of logging, the era of wanton industrial water pollution, and cities dumping untreated sewage into the Fraser River and its tributaries. The sockeye survived intense fishing pressure and a railway accident that dumped tons of rocks and debris into the Fraser, temporarily blocking the fishes' passage. Forestry rules have been tightened to protect rivers and streams, and the lower Adams River is now protected by Roderick Haig-Brown Provincial Park. Laws have been passed to crack down on industrial and municipal water pollution. Pulp and paper mills in Canada discharge 99 percent less dioxins and furans than they did in the past. Sewage treatment infrastructure has been upgraded. Today, the Adams River sockeye run is a gorgeous, living symbol of optimism.

Signs of hope and recovery are all around us. I've witnessed chum salmon spawning in Vancouver's Still Creek after more than eight decades of absence. Herring are spawning again in False Creek, right in the heart of Vancouver. Not that long ago, False Creek was an ugly urban wasteland degraded by a century of industrial activities. In recent years, condominiums, parks, playgrounds, and paths have replaced mills and factories. As the water has grown cleaner, native plants and wildlife have begun to return. An artificial island was built to provide habitat for a range of terrestrial and marine species. When a gray whale visited downtown Vancouver in 2010, swimming beneath the Burrard Bridge, it seemed like the whole city stopped. Word spread quickly on social media and thousands of people left their homes, schools, and jobs to come and witness this awe-inspiring sign of recovery. Vancouver is not alone. The Don River in the heart of Toronto is being restored. In downtown Seoul, Korea, a highway was removed so that Cheonggyecheon Stream could be restored. This exemplary project not only resulted in the return of mammals, fish, birds, amphibians, insects, and plants, but also decreased air pollution. In London, Stockholm, and New York, fish have returned to water bodies once choked with pollution. Nature's miraculous comeback stories are happening in virtually every country in the world.

THERE'S AN ENDURING MYTH that society must choose between environmental protection and economic well-being. But will going green really require costly sacrifices and decreases in our standard of living? No. This myth has been debunked by piles of academic studies, and more importantly by the experiences of communities, companies, and countries around the world. From Norway and Germany to Costa Rica and Brazil, from Vancouver to Stockholm, and from Alter Eco to Tesla, there is compelling evidence that the benefits of protecting the environment dwarf the costs. For example, while progress in reducing air pollution has cost industry billions of dollars over the past four decades, the total health and environmental benefits are measured in the trillions. Saving the ozone layer, securing access to safe drinking water, and eliminating the use of toxic chemicals have also resulted in tremendous net benefits. Germany's inspiring transition to renewable energy is costing households a paltry $280 per year, while the surge in green jobs drove unemployment to its lowest level since the reunification of East and West Germany in 1990.

Sweden, Norway, and Costa Rica are among the countries that are closest to achieving the bright green future depicted in this book. Norway is a major oil and gas exporter but imposes strict environmental rules on the industry and high taxes on pollution and profits. The impressive results include technological innovations, declining releases of toxic substances, and a publicly owned sovereign wealth fund worth more than one trillion dollars that will finance Norway's transition to a renewable, sustainable future. In Oslo, per capita greenhouse gas emissions were a mere 2.2 tonnes annually in 2014, 80–90 percent lower than those of the average North American. Sweden has cut greenhouse gas emissions by more than 20 percent since 1990, with most of the changes contributing to improved quality of life through cleaner air, safer streets, better public transit, healthier people, and more comfortable buildings. Both Norway and Sweden have recalibrated their economies by imposing taxes on pollution and waste, to reduce levels of these undesired items. Both countries consistently rank as the most generous in the world, dedicating 1 percent of their annual GDP to helping the world's poorest nations—four times the level of generosity provided by Canadian and American governments. Both these Scandinavian countries are discussing what a circular steady-state economy would look like, recognizing that today's levels of consumption in wealthy countries need to be reduced to alleviate pressure on overexploited planetary ecosystems.

In 2007, the Government of Sweden asked me to prepare an independent evaluation of their national environmental policies, programs, and progress. I presented my findings to the Riksdag, the Swedish Parliament, praising Swedish leadership on issues ranging from climate change to toxic substances. I also offered some constructive criticism regarding the need for Sweden to reduce their overall footprint, and warned that they were solving some of their environmental problems by exporting them to other countries. For example, reductions in logging in Swedish forests were offset by rising lumber and paper imports from countries whose forests may be more biodiverse. Declining oil use was offset by rising imports of biofuels from Brazil. To its credit, Sweden revised its goal of achieving sustainability within a generation to read "the overall goal of environmental policy [is] to hand over to the next generation a society in which the major environmental problems in Sweden have been solved, *and this should be done without increasing environmental and health problems outside Sweden's borders*" (emphasis added).

Costa Rica is widely recognized as an environmental leader as a result of decades of determined effort that includes the constitutional recognition of the right to a healthy environment; the enactment and implementation of strong environmental laws, such as prohibitions on open-pit mining and offshore oil and gas activity; placing more than one-quarter of its land in parks and protected areas, reversing the trend of deforestation; and producing 99 percent of its electricity from renewable energy. Twenty years ago, suggesting that Costa Rica's leading export would be anything but coffee or bananas would have provoked ridicule. Today, its most valuable export is computer chips, as high-tech giants have located manufacturing facilities to take advantage of the country's educated work force, clean air, and clean water. Costa Rica is the top-ranked country in the world on the Happy Planet Index, which integrates measures of life expectancy, self-rated happiness, and per capita ecological footprints. Costa Rica, Sweden, and Norway are not ecological utopias, but their successful pursuit of human well-being offers some inspiring lessons.

While the progress chronicled in this book is impressive, it is far from uniform. In most areas, Europe is ahead of North America. Many Latin American countries have strong environmental records despite lower levels of wealth. Asia and Africa lag behind. The pace of progress could be accelerated everywhere. Every country needs to recognize that all people, including future generations, have the right to live in a healthy environment. This right already enjoys constitutional protection in 110 countries but should be universally enshrined. Governments need to protect and fulfill this right by rigorously implementing and enforcing strong environmental laws and policies, using taxes and fees to make polluters pay, accelerating the shift to renewable energy, replacing environmental hazards with safe substitutes, applying the precautionary principle, and ensuring that environmental benefits and burdens are fairly shared. Making these changes is necessary to overcome the failures of the free market and curb the power wielded by multinational corporations. New jobs and business opportunities will be found in fields that protect or restore human and environmental health. As Paul Krugman wrote in 2014 in the *New York Times*, "If we ever get past the special interests and ideology that have blocked action to save the planet, we'll find that it's cheaper and easier than almost anyone imagines." Technology can make positive contributions but will not singlehandedly solve our problems. Old-fashioned ingenuity is required, and we have to be willing to change our ways. We need to cultivate more compassion, both for less fortunate humans and for the other species with whom we share the biosphere.

It is not only with respect to the environment that the world is making surprising strides. Ten years ago, I wrote an essay for the *Globe and Mail* about unsung progress on a host of issues fundamental to human well-being, including democracy, literacy, hunger, education, and health. An update of those findings provides additional reasons for optimism, but also raises red flags. One of the linchpins of a better future, democracy has made rapid strides in recent decades. Since 1990, the number of electoral democracies, where citizens freely choose their political leaders, has ballooned from 69 countries to 122 countries. Most of these nations have also ratified the world's major human-rights treaties, making a solemn commitment to respect, protect, and fulfill the rights of their citizens.

The UN's ambitious Millennium Development Goals anti-poverty campaign, launched in 2000, did not meet all of its 2015 targets but contributed to impressive progress. Like democracy, literacy is a cornerstone of a healthy, prosperous, and just society. The literacy rate in the developing world has jumped from 47 percent to 84 percent since 1970, meaning that billions of people are better equipped to boost their standard of living. The school enrolment rate in primary education in developing regions climbed from 83 to 90 percent between 2000 and 2012. Girls made the greatest gains, narrowing the gender gap.

Remarkable gains have been made in the efforts to address tuberculosis and malaria. Intensive efforts to overcome tuberculosis saved an estimated 22 million lives worldwide since 1995. Between 2000 and 2012, an estimated 3.3 million deaths from malaria were avoided due to increased antimalaria interventions. About 90 percent of those avoided deaths were children under the age of five. However, some of the main interventions include indoor pesticide spraying and insecticide-treated bed nets.

Child and maternal mortality have dropped dramatically. The number of kids who die before the age of five has been cut in half over the past 20 years. In developing countries, a child born today will live, on average, 25 years longer than a child born in 1950. Since 1990, the rate of maternal mortality (the number of women who die while giving birth) has also been reduced by almost 50 percent. Life expectancy in wealthy nations also continues to rise: in Canada, the average is 85 years for women and 80 years for men.

Although the world's population doubled between 1960 and 2000 and recently surpassed seven billion, the rate of growth is declining. Thanks to efforts to empower women and encourage family planning, families on every continent are now having fewer children. Globally, the average fertility rate has fallen from 6 children per woman to 2.4 in the past 50 years. Just one generation ago, the fertility rate in Thailand was 7, while today it is 1.6. In more than half of all nations, including Canada and China, the fertility rate is less than 2.1 children per woman. Populations in these countries will decline without immigration.

The world's poor are making economic progress. In 1990, almost half of the population in developing regions lived on less than $1.25 per day. The global rate of poverty has been sliced in half since then, reducing the number of people living in extreme poverty by more than 700 million. Between 1990 and 2012, average per capita income in developing countries, adjusted for inflation, doubled. In other words, people in the world's poorest nations have twice the buying power they had 25 years ago. Levels of human development have increased in all but 3 of 135 developing countries since 1970.

Dire predictions of widespread famine have not come true. Remarkably, global food production has grown at a faster rate than the population. Fewer people in total are malnourished today than in 1970. New UN data show that the number of undernourished people has fallen by more than 200 million since 1990, a tremendous achievement in light of population growth during that same time. Yet these gains involved extensive use of pesticides, fossil fuels, and fertilizers, and the replacement of wildlife habitat with land dedicated to agriculture.

The first Millennium Development Goal to be achieved was reducing by half the proportion of people who lack access to improved sources of water. Between 1990 and 2012, more than a billion people gained access to safe drinking water and almost two billion people gained access to improved sanitation facilities.

Another piece of surprisingly good news is the trend towards a more peaceful world. War, including wars between countries, civil wars, and colonial wars, has declined in frequency and intensity. Deaths caused by wars in the first 14 years of the 21st century averaged 55,000 per year, roughly half of the average during the 1990s (100,000 per year) and less than one-third the average during the Cold War from 1950 to 1990 (180,000 per year). The last decade saw fewer war deaths than any decade in the past century, according to Norway's Peace Research Institute in Oslo.

These extraordinary statistics tell only a tiny bit of the story. Behind the numbers are flesh and blood people living healthier, happier, longer lives than was possible in the past. The well-being of individuals, families, and communities has improved immeasurably, with more love and less loss. I'm not suggesting the world's problems are solved. Poverty persists. Thousands of children still die each and every day from preventable and treatable diseases. Hundreds of millions of people still lack access to electricity, sufficient food, safe drinking water, sanitation facilities, and health care. However, the progress that has been made should inspire the world to redouble its efforts in the post-2015 era.

But I must insert two major environmental caveats here, related to overconsumption and overpopulation. We can't just keep on biggering and biggering, as Dr. Seuss warned in his children's book *The Lorax*. Infinite growth on a finite planet is unsustainable by definition, violating the laws of physics. Even renewable energy, electric vehicles, cradle-to-cradle products, organic agriculture, and a circular economy have environmental impacts. While billions of people living in poverty need to consume more in order to reach a reasonable standard of living, the profligate consumption of people in the North needs to be moderated, recognizing that the pathway to health and happiness isn't through more stuff, but through family, friends, fulfilling work, spending time outside, and other aspects of life that don't involve consuming resources. The North American standard of living is too environmentally damaging to be enjoyed by seven billion people. The projected increase of two billion people by 2050 will make it harder to solve all of the world's environmental problems. Indeed, much or all of the progress outlined in this book could be overwhelmed by population growth. Tackling population means educating girls, empowering women, and improving access to family planning, contraception, and reproductive health services.

Despite the magnitude of these challenges, a sustainable future is possible. Whether we are able to grasp the opportunity really boils down to a matter of choices. More accurately, the collective outcome of billions of choices. Each of us has the potential to shift the balance. We can choose careers in a wide range of fields that support a green future, from art to engineering, and law to education. The prospects have never been better for employment that is rewarding, fulfilling, and environment-enhancing. Society needs green engineers to design smart grids, electric vehicles, geothermal power plants, wind turbines, and low-impact infrastructure. Tens of millions of green-collar jobs will be created to build them. Society needs farmers to produce healthy food close to markets, optimally with organic agricultural practices. We need workers to install millions of solar panels and solar hot water systems. We'll need generations of environmental educators, wilderness guides, and camp counsellors. There'll be more jobs in recycling and fewer in mining. There'll be more jobs in renewable energy and fewer in coal, oil, and natural gas. Public transit, green infrastructure, and sustainable agriculture create far more jobs per million dollars invested than their competitors. In terms of the big picture, there's compelling evidence that there will be less unemployment in a green economy, as Germany's experience illustrates. We need to be green consumers, buying less stuff and supporting environmentally and socially responsible businesses. Sharing goods can be less expensive and more convenient. We need to be green investors, ensuring that our savings are placed in institutions and investment vehicles that avoid or screen out corporations that plunder ecosystems and violate human rights. We can figure out ways to travel emissions-free or drive and fly less often. We can do renovations so that our homes produce more energy than they consume or at least substantially reduce the heating and cooling costs. We can eat meat less often, benefitting both our health and the planet.

Most importantly, we can be green citizens. We can exercise our democratic rights by participating enthusiastically in public debates, sharing our knowledge and values, and supporting political candidates who are ecologically literate and socially progressive. We can all lend our voices to the voiceless, including future generations and Nature. We need to be more like Sweden and Norway, where all parties offer strong environmental platforms and it's career limiting for politicians to take anti-environmental positions. I'm fortunate to live in the Saanich–Gulf Islands riding in British Columbia, represented in Ottawa by Elizabeth May, Canada's first Green Party Member of Parliament. In a fractious and hyperpartisan House of Commons, May has been a model politician, working tirelessly for her constituents and her country.

In recognition of May's remarkable efforts, her colleagues have voted her Parliamentarian of the Year, Canada's hardest-working MP, and best orator in Parliament. But the political systems in Canada and the U.S. need makeovers because they are producing undemocratic outcomes. In Canada, the archaic first-past-the-post system enabled a majority government to be elected in 2011 by only one-quarter of eligible voters. The Green Party earned 4 percent of the vote in 2011 but only 0.3 percent of the elected MPs (1 out of 308). Some form of proportional representation would be a superior system, ensuring that every vote counts. In the U.S., reforms are urgently needed to reduce the disproportionate influence of wealthy individuals and large corporations.

In a bright green future, everyone is ecologically literate, meaning that we all understand our fundamental dependence on natural systems for the air we breathe, the water we drink, the food we eat, the medicines that heal our injuries and illnesses, and many of the materials used to make everything from clothing to housing. Today, far too many people are clueless about where their tap water comes from; where their wastes go when they're put in the garbage, recycling bin, or toilet; which plants and animals are native versus introduced or invasive; or how their electricity is generated. Most individuals underestimate the magnitude of their own environmental impacts. Environmental education needs to be a core element of the curriculum from kindergarten through university, on par with reading, writing, and arithmetic. Ecologically literate citizens are better equipped to hold politicians accountable for environmental progress, see through corporate green-washing, and ensure that their communities and countries are moving in the right direction.

A vital element of environmental literacy is simply spending time outside in natural settings. Research proves that this has physical and psychological benefits. Nature's wonders can be found everywhere, not just in the Galapagos Islands or the Serengeti. Even in cities there are green oases, from parks to apartment balconies, where plants are growing, birds are nesting, and wildlife is thriving. For example, millions of migratory shorebirds stop in the Fraser River estuary on the outskirts of Vancouver to refuel for their incredible journeys to the Arctic in the spring and to the tropics in the fall. We've wandered out onto the mud flats at the height of this migration, surrounded by hundreds of thousands of sandpipers twittering away while they feast on tiny mud shrimp. Occasionally a raptor will pass overhead and the sandpipers will interrupt their meal in a spontaneous but seemingly choreographed flight, sunlight reflecting off their bodies. At marvellous moments like these, you feel completely connected to, integrated with, the rest of Nature.

The life on this beautiful planet is a miracle of the highest order. Scientists have scoured the universe for signs of life and come up empty. The elements, processes, and location of this planet may be a billion- or even trillion-to-one long shot. Every astronaut who has orbited Earth has been gobsmacked by the beauty and fragility of this blue dot. Our gratitude ought to be boundless, and so should our efforts to take care of this place.

As I finished writing this chapter, a great blue heron unleashed a pterodactyl-like squawk, flying past with languid flaps of its super-sized wings. The heron landed atop a tall Douglas fir tree, borrowing the perch often held by bald eagles on the islet to the west. From there, it surveyed James Bay for its breakfast. I'm delighted that we share this corner of the planet with such an elegant creature. The process of researching and writing this book has given me unprecedented confidence that we are, contrary to conventional wisdom, both capable of and on track to achieve a much brighter and greener future than most people would predict. I've regained my sense of optimism. Based on what I've learned, I'm inspired to take additional actions. We'll be installing solar panels on the roof of our house that will eliminate our electricity bill and even generate a bit of income. I've joined a group that's raising money to put a solar photovoltaic system of the roof of every suitable public building in our community by 2020. We secured a $20,000 grant from the Nu-to-Yu (the volunteer-run thrift store that has donated over $1 million to local charities) that enabled us to complete our first project (a 10 killowatt system composed of 39 solar panels) at Pender Island's popular recycling depot. Our family's next car will be fully electric, as we're already looking into trading our Toyota Prius for a Nissan Leaf. I've promised Meredith that we'll go on a sea-kayaking trip to the west coast of Vancouver Island so we can witness and celebrate the recovery of sea otter populations first hand.

There are some things that unite people, regardless of age, colour, nationality, gender, religion, or political beliefs: We all want to live in places that are vibrant, beautiful, affordable, and sustainable. We all want tomorrow to be at least as good as, and preferably better than, today. We all want to live in an environment that's healthy—to breathe clean air, drink safe water, eat delicious, healthy food, not have to worry about toxic chemicals, and know that biodiversity is flourishing. And we want to be sure that our children and grandchildren will enjoy a world that is just as wonderful as the world we've experienced.

The belief that something positive is possible is an essential step towards making it happen. This book demonstrates that we have the capacity to overcome today's environmental challenges in ways that will make a better tomorrow. We can achieve a world with far less pollution, powered by 100 percent renewable energy from sun, wind, and water. Skies will be blue more often than brown. There'll be more birdsong and frog choruses, fewer sirens and less machine noise. Cities will have less pavement and we'll all live within a five-minute walk of public green spaces—parks, community gardens, and orchards. Water will be clean instead of contaminated. Millions of premature deaths, countless preventable illnesses, and billions of dollars in unnecessary healthcare expenses will be avoided by reducing pollution. People will be healthier and happier. Words like waste, garbage, and pollution will become anachronisms. The optimistic environmentalist's message to the world is that we can do better. The solutions to all of our problems are readily available, and implementing them will make us wealthier, not poorer. From Vancouver to Stockholm, this vision is becoming a reality.

Success breeds success. Every green business, green building, improvement in transit service, solar PV system or hot water heater, electric vehicle, trip made on foot or by bike, new community garden or orchard, and building that produces more energy than it consumes will generate additional momentum. Each endangered species that recovers to a healthy population, newly designated protected area, organic farm, household connected to safe drinking water and sanitation, cradle-to-cradle product, and innovation inspired by Nature paves the way for additional action. A virtuous circle has begun, and eventually a tipping point will be reached.

We know in our very bones the vital importance of shifting to a more sustainable way of being. Through optimism, belief, and effort, we can and will get there. Within the geologically infinitesimal span of one or two generations—ours and our children's—we can ensure a cleaner, greener, healthier, and happier future for all of Earth's inhabitants.

"I am an optimist. It does not seem too much use being anything else."
—Winston Churchill

Selected Bibliography

The following sources provide much of the background information for this book. For specific citations, please contact the author at optimisticenvironmentalist@gmail.com.

INTRODUCTION: THE IMPORTANCE OF BEING OPTIMISTIC

Carver, C.S., and M.F. Scheier. 2014. "Dispositional Optimism." *Trends in Cognitive Sciences* 18 (6): 293–299.

Conversano, C., A. Rotondo, E. Lensi, et al. 2010. "Optimism and Its Impact on Mental and Physical Well-Being." *Clinical Practice and Epidemiology in Mental Health* 6: 25–29.

Heller, K. 2012. "Depression in Teens and Children." *Psychology Central*, psychcentral.com/lib/depression-in-teens-and-children/00010763.

International Union for the Conservation of Nature. 2015. *The IUCN Red List*, iucnredlist.org.

Leopold, A. 1949. *A Sand County Almanac: With Essays on Conservation from Round River.* New York: Oxford University Press.

Monbiot, G. 16 June 2014. "An Ounce of Hope Is Worth a Ton of Despair," monbiot.com/2014/06/16/an-ounce-of-hope-is-worth-a-ton-of-despair.

Seligman, M.E.P. 2006. *Learned Optimism: How to Change Your Mind and Your Life* (2nd ed.). New York: Vintage.

Sharot, T. 2011. *The Optimism Bias: A Tour of the Irrationally Positive Brain*. New York: Pantheon. See also Tali Sharot's Ted Talk on optimism in February 2012, ted.com/talks/tali_sharot_the_optimism_bias?language=en.

Steptoe, A., C. Wright, S.R. Kunz-Ebrecht, et al. 2006. "Dispositional Optimism and Health Behavior in Community-Dwelling Older People: Associations with Healthy Aging." *British Journal of Health Psychology* 11: 71–84.

Tindle, H. 2013. *Up: How Positive Outlook Can Transform Our Health and Aging*. New York: Penguin.

1. Nature's Comeback Stories

Barkham, P. 7 April 2014. "Endangered Butterfly Defies Climate Change with New Diet and Habitat." *The Guardian*.

Boucher, D., P. Elias, J. Faires, and S. Smith. 2014. *Deforestation Success Stories: Tropical Nations Where Forest Protection and Reforestation Policies Have Worked*, ucsusa.org/sites/default/files/legacy/assets/documents/global_warming/deforestation-success-stories-2014.pdf.

Goodall, J. 2009. *Hope for Animals and Their World: How Endangered Species Are Being Rescued from the Brink*. London: Grand Central.

International Court of Justice. 2014. *Judgment: Whaling in the Antarctic (Australia v. Japan, New Zealand intervening)*, icj-cij.org/docket/files/148/18136.pdf.

International Union for the Conservation of Nature. 2014. *The IUCN Red List: Celebrating 50 Years of Conservation*. Washington: IUCN.

International Union for the Conservation of Nature. 2015. *World Database on Protected Areas*, iucn.org/about/work/programmes/gpap_home/

gpap_biodiversity/gpap_wdpa.

MacKinnon, J.B. 2013. *The Once and Future World: Nature as It Was, as It Is, as It Could Be*. Toronto: Vintage Canada.

Manning, R. 2011. *Rewilding the West: Restoration in a Prairie Landscape*. Los Angeles: University of California Press.

Monbiot, G. 2013. *Feral: Rewilding the Land, the Sea, and Human Life*. London: Allen Lane.

Nepstad, D., D. McGrath, C. Stickler, et al. 2014. "Slowing Amazon Deforestation through Public Policy and Interventions in Beef and Soy Supply Chains." *Science* 344 (6188): 1118–1123.

Perrin, W.F., B. Wursig, and J.G.M. Thewissen. 2009. *Encyclopedia of Marine Mammals* (2nd ed.). Academic Press.

Suckling, K., N. Greenwald, and T. Curry. 2012. *On Time, On Target: How the Endangered Species Act Is Saving America's Wildlife*. Center for Biological Diversity.

U.S. Fish and Wildlife Service. 2007. *Bald Eagles: Life Story and Conservation Success*, fws.gov/midwest/Eagle/recovery/index.html.

U.S. Fish and Wildlife Service. *Recovery Success Stories*, fws.gov/endangered/what-we-do/recovery-stories.html.

World Wildlife Fund, Zoological Society of London, and Global Footprint Network. 2012. *Living Planet Report 2012*. London: WWF.

2. The Renewable Energy Revolution

Bradford, T. 2006. *The Solar Revolution: The Economic Transformation of the Global Energy Industry*. Cambridge, MA: MIT Press.

Canadian Wind Energy Association, canwea.ca.

Crook, E. 18 September 2014. "U.S. Sun and Wind Start to Outshine Gas." *Financial Times*.

Delucchi, M.A., and M.Z. Jacobson. 2011. "Providing All Global Energy with Wind, Water, and Solar Power, Part II: Reliability, System and Transmission Costs, and Policies." *Energy Policy* 39: 1170–1190.

Editorial. 17 January 2015. "Seize the Day: The Fall in the Price of Oil and Gas Provides a Once-in-a-Generation Opportunity to Fix Bad Energy Policies." *The Economist*.

Electricity Human Resources Canada. 2013. *A National Human Resources Strategy for Renewable Electricity*, renewingfutures.ca/CMFiles//EHRCrfs.pdf.

Gillis, J. 13 September 2014. "Sun and Wind Alter Global Landscape, Leaving Utilities Behind." *New York Times*.

Gladwell, M. 2002. *The Tipping Point: How Little Things Can Make a Big Difference*. New York: Little, Brown and Company.

Global Wind Energy Council, gwec.net.

International Renewable Energy Agency. 2014. *Renewable Power Generation Costs in 2014*. Abu Dhabi: IRENA.

International Energy Agency. 2014. *Energy Technology Perspectives: Harnessing Electricity's Potential*. Paris: IEA, iea.org/Textbase/npsum/ETP2014SUM.pdf.

International Energy Agency. 2014. *World Energy Outlook 2014*. Paris: IEA, iea.org/Textbase/npsum/WEO2014SUM.pdf.

Jacobson, M.Z., and M.A. Delucchi. 2011. "Providing All Global Energy with Wind, Water, and Solar Power, Part I: Technologies, Energy Resources, Quantities and Areas of Infrastructure, and Materials." *Energy Policy* 39: 1154–1169.

Kennedy, D. 2012. *Rooftop Revolution: How Solar Power Can Save Our Economy—and Our Planet—From Dirty Energy*. San Francisco: Berrett-Koehler.

Leggett, J., ed. 2009. *The Solar Century: The Past, Present, and World-Changing Future of Solar Energy*. London: Green Profile.

Natural Resources Canada. "About Electricity," nrcan.gc.ca/energy/
electricity-infrastructure/about-electricity/7359.

Shell. *New Lens Scenarios*, shell.com/global/future-energy/scenarios.html.

Sierra Club. *Beyond Coal: Victories*, content.sierraclub.org/coal/victories.

Torrie, R.D., T. Bryant, M. Beer, et al. 2013. *An Inventory of Low-Carbon Energy for Canada*. Vancouver: Trottier Energy Futures Project.

3. The Circular Economy

Anielski, M. 2007. *The Economics of Happiness: Building Genuine Wealth*. Gabriola Island: New Society Publishers.

AskNature online library, asknature.org.

Benyus, J. 1997. *Biomimicry: Inspiration Inspired by Nature*. New York: William Morrow.

Biomimicry Institute, biomimicry.org.

Boulding, K. 1966. The Economics of the Coming Spaceship Earth, eoearth.org/view/article/156525.

Cradle to Cradle Products Innovation Institute, c2ccertified.org.

Ellen McArthur Foundation. Circular Economy Initiative, ellenmacarthurfoundation.org/circular-economy.

European Commission. 2014. Moving Towards a Circular Economy, ec.europa.eu/environment/circular-economy.

Hawken, P., A. Lovins, and L.H. Lovins. 1999. *Natural Capitalism: Creating the Next Industrial Revolution*. Boston: Little, Brown and Company.

McDonough, W., and M. Braungart. 2002. *Cradle to Cradle: Remaking the Way We Make Things*. New York: North Point Press.

McDonough, W., and M. Braungart. 2013. *The Upcycle: Beyond Sustainability—Designing for Abundance*. New York: North Point Press.

4. Cleaner Air

Bates, D.V., and R.B. Caton, eds. 2002. *A Citizen's Guide to Air Pollution* (2nd ed.). Vancouver: David Suzuki Foundation.

Bienkowski, B. 6 March 2014. "Soot Success: Clean Air Within Reach Nationwide, But Not for Long," *Environmental Health News*.

Canadian Medical Association. 2008. *No Breathing Room: National Illness Costs of Air Pollution*. Toronto: CMA.

Doyle, J. 2000. *Taken for a Ride: Detroit's Big Three and the Politics of Pollution*. New York: Four Walls, Eight Windows.

European Environment Agency. 19 November 2014. "European Urbanites Breathing Highly Polluted Air," eubusiness.com/news-eu/environment-health.yso.

Jessiman, B., and R. Burnett. 2002. "Sulphur in Gasoline." *Health and the Environment: Critical Pathways* (4). Ottawa: Health Canada.

O'Connor, A.M. 11 April 2010. "Mexico City Drastically Reduced Air Pollutants Since 1990s." *Washington Post*.

Reeves, F. 2014. *Planet Heart: How an Unhealthy Environment Leads to Heart Disease*. Vancouver: Greystone.

Taylor, E., and A. McMillan, eds. 2013. *Air Quality Management: Canadian Perspectives on A Global Issue*. London: Springer.

U.S. Environmental Protection Agency. 2011. *The Benefits and Costs of the Clean Air Act from 1990 to 2020. Final Report*. Washington: EPA.

U.S. Environmental Protection Agency. *The Benefits and Costs of the Clean Air Act, 1970 to 1990*. Washington: EPA.

U.S. Environmental Protection Agency. "Progress Cleaning the Air and Protecting People's Health," epa.gov/air/caa/progress.html.

5. Saving the Ozone Layer

Benedick, R.E. 1991. *Ozone Diplomacy: New Directions in Safeguarding the Planet.* Cambridge, MA: Harvard University Press.

Boyd, D.R. 2003. *Unnatural Law: Rethinking Canadian Environmental Law and Policy.* Vancouver: UBC Press.

de Gruijl, F.R., and J.C. van der Leun. 2000. "Environment and Health: Ozone depletion and ultraviolet radiation." *Canadian Medical Association Journal* 163 (7): 851–855.

Molina, M.J., and F.S. Rowland. 1974. "Stratospheric Sink for Chlorofluoromethanes: Chlorine Atom Catalyzed Distribution of Ozone." *Nature* 249: 810–812.

NASA. 2012. *From Discovery, to Solution, to Evolution: Observing Earth's Ozone Layer,* nasa.gov/topics/earth/features/ozone-history.html.

Newman, P.A., L.D. Oman, A.R. Douglass, et al. 2009. "What Would Have Happened to the Ozone Layer if Chlorofluorocarbons (CFCs) Had Not Been Regulated?" *Atmospheric Chemistry and Physics* 9: 2113–2128.

Suzuki, D., and A. Gordon. 1990. *It's a Matter of Survival.* Toronto: Stoddart.

UN Environment Programme and World Meteorological Organization. 2014. *Ozone Layer on Track to Recovery: Success Story Should Encourage Action on Climate.* Joint Press Release, 10 September 2014. Geneva: UNEP/WMO.

U.S. Environmental Protection Agency. 2007. *Achievements in Stratospheric Ozone Protection: Progress Report.* Washington: EPA.

6. Taps, Toilets, and Farms

Baranski, M., D. Srednicka-Tober, N. Volakakis, et al. 2014. "Higher Antioxidant and Lower Cadmium Concentrations and Lower Incidence of Pesticide Residues in Organically Grown Crops: A Systematic Literature Review and Meta-Analyses." *British Journal of Nutrition* 112 (5): 794–811.

Barlow, M. 2013. *Blue Future: Protecting Water for People and the Planet Forever.* Toronto: Anansi.

Boyd, D.R. 2011. "No Taps, No Toilets: First Nations and the Constitutional Right to Water in Canada." *McGill Law Journal* 57 (1): 81–134.

Fair Trade International, fairtrade.net.

Goodall, J. 2005. *Harvest for Hope: A Guide to Mindful Eating.* New York: Warner Wellness.

Griffith-Greene, M. 8 March 2014. "Pesticide Traces in Some Tea Exceed Allowable Limits." *CBC News.* cbc.ca/news/canada/pesticide-traces-in-some-tea-exceed-allowable-limits-1.2564624.

Lappe, F.M., and A. Lappe. 2003. *Hope's Edge: The Next Diet for a Small Planet.* New York: Tarcher.

Melina, V., and B. Davis. 2003. *Becoming Vegetarian: The Complete Guide to Adopting a Healthy Vegetarian Diet* (2nd ed.). Toronto: Wiley.

Nestle, M. 2006. *What to Eat.* New York: North Point Press.

Pollan, M. 2008. *In Defense of Food: An Eater's Manifesto.* New York: Penguin.

Sandford, R.W. 2009. *Restoring the Flow: Confronting the World's Water Woes.* Canmore, AB: Rocky Mountain Books.

Smith, A., and J.B. Mackinnon. 2007. *The 100-Mile Diet: A Year of Local Eating.* Toronto: Random House.

UN Millennium Development Goals, un.org/millenniumgoals.

UN Special Rapporteur on the Human Right to Safe Drinking Water and Sanitation, sr-watersanitation.ohchr.org.

7. Global Detox

Basel Convention on the Control of Transboundary Movements of Hazardous Wastes and their Disposal, basel.int.

Bohme, S.R. 2014. *Toxic Injustice: A Transnational History of Exposure and Struggle.* Los Angeles: University of California Press.

Boyd, D.R. 2015. *Cleaner, Greener, Healthier: A Prescription for Stronger Environmental Laws in Canada*. Vancouver: UBC Press.

Branchi, I., F. Capone, E. Alleva, et al. 2003. "Polybrominated Diphenyl Ethers: Neurobehavioral Effects Following Developmental Exposure." *Neurotoxicology* 24 (3): 449–462.

Darnerud, P.O. 2003. "Toxic Effects of Brominated Flame Retardants in Man and Wildlife." *Environment International* 29 (6): 841–853.

Draxler, B. 15 January 2014. "Female Sea Snails No Longer Growing Penises Thanks to Ban on Toxic Chemical." *Discover Magazine*.

Hagerty, J.R. 16 December 2013. "California Judge Orders Lead Paint Cleanup." *Wall Street Journal*.

Health Canada. 2013. *Second Report on Human Exposure to Environmental Chemicals in Canada*, hc-sc.gc.ca/ewh-semt/pubs/contaminants/chms-ecms-cycle2/overview-vue-eng.php.

Herbstman J.B., A. Sjodin, M. Kurzon, et al. 2010. "Prenatal Exposure to PBDEs and Neurodevelopment." *Environmental Health Perspectives* 118 (5): 712–719.

Michaels, D. 2009. *Doubt Is Their Product: How Industry's Assault on Science Threatens Your Health*. New York: Oxford University Press.

Minimata Convention on Mercury, mercuryconvention.org.

Rotterdam Convention on the Prior Informed Consent Procedure for Certain Hazardous Chemicals and Pesticides in International Trade, pic.int.

Smith, R., and B. Lourie. 2009. *Slow Death by Rubber Duck: How the Toxic Chemistry of Everyday Life Affects Our Health*. Toronto: Knopf.

Stockholm Convention on Persistent Organic Pollutants, pops.int.

U.S. Centers for Disease Control and Prevention. 2014. *National Report on Human Exposure to Environmental Chemicals, Fourth Report*, cdc.gov/exposurereport/index.html.

Wolf, L.K. 2014. "The Crimes of Lead." *Chemical and Engineering News* 92 (5): 27–29.

8. The Greenest City Decathlon

Boyd, D.R. 2010. *Vancouver 2020: A Bright Green Future. An Action Plan for Becoming the World's Greenest City.* City of Vancouver: Greenest City Action Team.

C-40 Cities Climate Leadership Group, c40.org/cities.

City Climate Leadership Awards. cityclimateleadershipawards.com.

European Commission. *European Green Capital Award,* ec.europa.eu/environment/europeangreencapital/index.html.

Gondor, D., and T. Hamilton. 2013. "Sustainable Cities." *Corporate Knights.* 52–67.

Grassroots Recycling Network, grrn.org.

Montgomery, C. 2013. *Happy City: Transforming Our Lives Through Urban Design.* Toronto: Doubleday.

Siemens. 2015. Green City Index, siemens.com/entry/cc/en/greencityindex.htm.

Speck, J. 2012. *Walkable City: How Downtown Can Save America, One Step at a Time.* New York: North Point Press.

Stockholm official website, international.stockholm.se.

Vancouver. 2011. *Vancouver 2020 Greenest City Action Plan,* vancouver.ca/green-vancouver/a-bright-green-future.aspx.

9. The Future of Buildings

Bernhardt's Passive Home (Victoria), bernhardtpassive.com.

California Public Utilities Commission. 2013. *Residential Zero Net Energy Action Plan,* californiaznehomes.com.

Consumer Reports. 2006. *Complete Guide to Reducing Energy Costs.* Yonkers, NY: Consumer's Union of the United States.

European Commission. *Nearly Zero Energy Buildings,* ec.europa.eu/energy/en/topics/energy-efficiency/buildings/nearly-zero-energy-buildings.

Green Energy Futures. 2014. "Chasing Net-Zero" (episodes 78–81 and 103), greenenergyfutures.ca/episodes.

Keenan, S. 14 August 2013. "The Passive House: Sealed for Freshness," *New York Times*.

Living Building Challenge, living-future.org/lbc.

Stoyke, S. 2007. *The Carbon Buster's Home Energy Handbook: Slowing Climate Change and Saving Money*. Gabriola Island: New Society Publishers.

UniverCity Childcare. 2013. *Sustainable Architect and Building Magazine*, sabmagazine.com/blog/2013/06/10/univercity-childcare-burnaby-bc.

U.S. Department of Energy. 2006. *Energy Saver$: Tips on Saving Energy and Money at Home*. Washington: DOE.

Visitor Centre, VanDusen Botanical Garden (Vancouver), vandusengarden.org/explore/vandusen-botanical-garden/visitor-centre-discovery-hall.

10. Electrifying Transport

Clean Technica (Transport), cleantechnica.com/category/clean-transport-2.

Hopkins, D. 2013. "Changing Youth Mobility Practices: Exploring Transport Modal Choice and Driver's Licensing Behaviours Among Adolescents and Young Adults," academia.edu/6022934/Changing_Youth_Mobility_Practices.

Inside Electric Vehicles, insideevs.com.

MacKay, D.C. 2009. *Sustainable Energy: Without the Hot Air*. UIT Cambridge Ltd., withouthotair.com.

Madigan, K. 21 January 2014. "Vital Signs: More Households Don't Own a Car," *Wall Street Journal*, blogs.wsj.com/economics/2014/01/21/vital-signs-more-households-dont-own-a-car.

McCredie, A. 24 July 2014. "Hybrid Highway: Trail-Blazing Cabbie." *The Province*, blogs.theprovince.com/2014/07/24/hybrid-highway-trial-blazing-cabbie.

Rocky Mountain Institute (Amory Lovins), rmi.org/transportation.

Sun Country Highway, suncountryhighway.ca.

Tesla Motors, teslamotors.com.

CONCLUSION: FROM OPTIMISM TO ACTION

Boyd, D.R. 2015. *Cleaner, Greener, Healthier: A Prescription for Stronger Environmental Laws in Canada*. Vancouver: UBC Press.

New Economics Foundation. 2012. *The Happy Planet Index 2012 Report: A Global Index of Sustainable Well-Being*. London: NEF.

Peace Research Institute Oslo. 2014. *Data on Armed Conflicts*, prio.org/Data/Armed-Conflict/?id=348.

Seligman, M.E.P. 2006. *Learned Optimism: How to Change Your Mind and Your Life* (2nd ed.). New York: Vintage.

Seligman, M.E.P. 2007. *How to Raise an Optimistic Child: A Proven Program to Safeguard Children Against Depression and Build Lifelong Resilience*. New York: Mariner.

Swedish Environmental Protection Agency. 2011. *Swedish Consumption and the Environment*. Stockholm: SEPA.

Suzuki, D., and H. Dressel. 2002. *Good News for a Change: How Everyday People Are Helping the Planet*. Vancouver: Greystone.

Turner, C. 2007. *The Geography of Hope: A Tour of the World We Need*. Toronto: Random House.

UN Millennium Development Goals, www.un.org/millenniumgoals.

Acknowledgements

IT TAKES A COMMUNITY TO WRITE A BOOK, and I'm grateful for the support, inspiration, and encouragement of Margot and Meredith, Maude Barlow, Mark Bernhardt, Kate Biddell, Alaya Boisvert, Bruce Boland, Sandy, Jeanette, Sonje, and Seamus Boyd, Silver Donald Cameron, Mike Cormack, Mary Cottrell, Jenny Drake, Denny Goertz, Joanne Green, Lisa Fleming, Mark Haddock, Don Harrison, Will Horter, Ian Kean, Deb Lacroix, Elizabeth May, John, Stan, and Liz Mills, Carol Newell, David Ohnona, Devon Page, Lynda Prince, Andrea Reimer, Ken Rempel, Susan Renouf, Paul Richardson, Gregor Robertson, Mary, Darwin, and Nicola Schatz, Ethan Smith, Joel Solomon, David Suzuki, Anne and Peter Venton, Scott Wallace, and Nia Williams.

This is my first book with ECW Press, and from start to finish it has been a tremendous experience. Susan Renouf generously shared her wisdom and insights. From our very first conversation, we connected like old friends. Erin Creasey, Jen Knoch, Crissy Calhoun, Michel Vrana, and the rest of the ECW team have done an outstanding job, and I look forward to working together on many more books.

I would also like to thank Canada, the beautiful country I am so fortunate to call home. As a Canadian who has experienced many of our incredible natural wonders, maybe it's easy for me to be optimistic. I've paddled down mind-blowingly gorgeous rivers that most people have never even heard of, including the Whiting, Turnagain, and Ravens Throat. I grew up hiking and cross-country skiing at Banff, Kananaskis, and Waterton in the Rocky Mountains. My brother and I once mountaineered for three weeks from the Whistler ski resort to the Pacific Ocean, traversing glaciers and icefields, seeing virtually no evidence of human impacts. Despite all the damage that's been done, Canada is still one of the wildest countries in the world.

We live in a community (Pender Island) where kids feel at home on the beach or in the forest. Children love spending time outdoors in natural settings. They thrive and develop deep connections to Nature. We've also taken Meredith to see some of Canada's natural treasures, including Long Beach near Tofino, Lake Louise (including the awesome hike up to the Plain of Six Glaciers teahouse), Takkakaw Falls, the Adams River, Jumbo Pass, the Bruce Peninsula, and Cathedral Provincial Park. One of my greatest hopes is that these experiences, and this book, will help Meredith become an optimistic environmentalist too.

About the Author

DAVID R. BOYD is an environmental lawyer, professor, and advocate for recognition of the right to live in a healthy environment. Boyd is the award-winning author of seven books and more than 100 articles and currently co-chairs Vancouver's Greenest City initiative with Mayor Gregor Robertson. He lives on Pender Island, B.C. For more information, visit DavidRichardBoyd.com.

A portion of the royalties of this book will be donated to the David Suzuki Foundation and to Ecojustice for their valuable work in championing everyone's right to live in a healthy environment.

This book is printed on FSC® certified paper.

Copyright © David R. Boyd, 2015

Published by ECW Press
665 Gerrard Street East, Toronto, Ontario, Canada M4M 1Y2
416-694-3348 / info@ecwpress.com

All rights reserved. No part of this publication may be reproduced, stored in a retrieval system, or transmitted in any form by any process — electronic, mechanical, photocopying, recording, or otherwise — without the prior written permission of the copyright owners and ECW Press. The scanning, uploading, and distribution of this book via the internet or via any other means without the permission of the publisher is illegal and punishable by law. Please purchase only authorized electronic editions, and do not participate in or encourage electronic piracy of copyrighted materials. Your support of the author's rights is appreciated.

Library and Archives Canada Cataloguing in Publication

Boyd, David R. (David Richard), 1964-, author
The optimistic environmentalist : progressing towards a greener future / David R. Boyd.

ISBN 978-1-77041-238-5
also issued as: 978-1-77090-763-8 (PDF);
978-1-77090-764-5 (epub)

Includes bibliographical references.
Issued in print and electronic formats.

1. Environmental policy. 2. Environmental protection. 3. Sustainable development. 4. Environmentalism. I. Title.

GE170.B69 2015   363.7   C2015-902763-2 | C2015-902764-0

Editor for the press: Susan Renouf
Cover design: Michel Vrana
Cover image: © Masa Ushioda/
CoolWaterPhoto.com

Author photo: Davy Rippner
Type: Rachel Ironstone

Printing: Marquis   5   4

The publication of *The Optimistic Environmentalist* has been generously supported by the Canada Council for the Arts, which last year invested $157 million to bring the arts to Canadians throughout the country. We acknowledge the support of the Ontario Arts Council (OAC), an agency of the Government of Ontario, which last year funded 1,793 individual artists and 1,076 organizations in 232 communities across Ontario, for a total of $52.1 million. We also acknowledge the financial support of the Government of Canada through the Canada Book Fund for our publishing activities, and the contribution of the Government of Ontario through the Ontario Book Publishing Tax Credit and the Ontario Media Development Corporation.

Printed and bound in Canada

Get the eBook
FREE!

At ECW Press, we want you to enjoy this book in whatever format you like, whenever you like.
Leave your print book at home and take the eBook to go!
Purchase the print edition and receive the eBook free.
Just send an email to ebook@ecwpress.com and include:

- the book title
- the name of the store where you purchased it
- your receipt number
- your preference of file type: PDF or ePub?

A real person will respond to your email with your eBook attached. Thank you for supporting an independently owned Canadian publisher with your purchase!